NARROW BAND-PASS FILTERS FOR LOW FREQUENCY APPLICATIONS

Evaluation of Eight Electronics Filter Design Topologies

Dr. Raman K. Attri

ISBN: 978-981-11-9872-4 (e-book)
ISBN 978-981-14-0133-6 (paperback)
First published: 2018
Draft written: 2005
Lead author: Raman K. Attri
Published by Speed To Proficiency Research: S2Pro©
Published at Singapore
Printed in the United States of America

National Library Board, Singapore Cataloguing in Publication Data

Name(s): Attri, Raman K., 1973-
Title: Narrow band-pass filters for low frequency applications : evaluation of eight electronics filter design topologies / Dr Raman K. Attri.
Other title(s): R. Attri instrumentation design series (electronics)
Description: Singapore : Speed To Proficiency Research, 2018. | Includes bibliographical references.
Identifier(s): OCN 1077481515 | ISBN 978-981-14-0133-6 (paperback) | ISBN 978-981-11-9872-4 (e-book)
Subject(s): LCSH: Electric network topology. | Electric filters, Bandpass. | Electronic books.
Classification: DDC 621.38215--dc23

Speed To Proficiency
RESEARCH

Speed To Proficiency Research: S2Pro©
A research and consulting forum
Singapore 560463
https://www.speedtoproficiency.com
rkattri@speedtoproficiency.com

To my little brother – Deepak, whose childhood I missed experiencing

R. Attri Instrumentation Design Series
(Electronics)

CONTENTS

ABOUT THE BOOK

Narrow Band-pass filtering techniques have been a challenging task since the inception of audio and telecommunication applications. The challenge involves keeping quality factor, gain and mid-frequency of the filter independent of each other. The critical applications require a design that ensures mid-frequency immune to the circuit component tolerances. It becomes increasingly difficult for low-frequency applications where the shift in few Hz in mid-frequency would cause desired frequencies to fall outside the filter's bandwidth and go undetected. The selection of right topology of the filter for the best performance is the key to successful design. This book objectively compares the relative performance of eight popular narrow band-pass filter topologies. The filter topologies are evaluated using a real-world practical example of designing an extremely narrow band-pass filter. The book provides guidelines for selecting the right topology for the low-frequency narrow band-pass filter.

PREVIOUS WORK

This book was originally conceptualized and authored in 2005-2006. As such it should be read remembering the time frame. The author does not claim contemporariness of the concepts, though the principles discussed in this book are universally applicable for analog electronics design. The chapters in this series previously appeared as working papers:

Attri, RK 2005, 'Evaluation of Single op-amp Topologies For Extremely Narrow Band-Pass Filter Design,' *R.Attri Instrumentation Design Series*

(*Electronics*), Paper No. 3, Sept 2005. https://dx.doi.org/10.13140/RG.2.1.4755.5920.

Attri, RK 2005, 'Design Analysis and Evaluation of 1 and 2 op-amp Topology for Design of Stable Narrow Band-Pass Filter,' *R.Attri Instrumentation Design Series (Electronics)*, Paper No. 5, Sept 2005. https://dx.doi.org/10.13140/RG.2.1.4231.3040.

Attri, RK 2005, 'Design of stable Narrow Band-Pass Filter using Multi-stage Biquad Topology,' *R.Attri Instrumentation Design Series (Electronics)*, Paper No. 4, Sept 2005. https://dx.doi.org/10.13140/RG.2.1.4493.4481.

Attri, RK 2005, 'Practical Design Evaluation of Extremely Narrow Band-Pass Filter Topologies,' *R.Attri Instrumentation Design Series (Electronics)*, Paper No. 7, Sept 2005. https://dx.doi.org/10.13140/RG.2.1.1872.0081.

Attri, RK 1998, 'Various Noise Sources & Noise Reduction Techniques in Instrumentation,' *R.Attri Instrumentation Design Series (Electronics)*, Paper No. 1, June 1998. https://dx.doi.org/10.13140/RG.2.1.2592.9046.

SUGGESTED CITATION

Attri, RK 2018, *Narrow Band-Pass Filters for Low-Frequency Applications: Evaluation of Eight Electronics Filter Design Topologies*, R.Attri Instrumentation Design Series (Electronics), ISBN 978-981-14-0133-6 (paperback), ISBN 978-981-11-9872-4 (e-book),Speed To Proficiency Research: S2Pro©, Singapore.

CHAPTER 1 – DESIGN REQUIREMENTS FOR EXTREMELY NARROW BAND-PASS FILTERS

1.1. Practical Challenges

Band-pass filter design has nevertheless been a challenge given many interrelated dependencies in the circuit parameters. In Band-pass filter, Quality Factor (Q) and gain of the filter are generally interrelated and thus do not give the independent control. There are always some design trade-offs that need to be made. In the case of the narrow band-pass filter, the circuit stability poses difficult requirements. Generally, the narrow band-pass filtering action is achieved by increasing the Q value of the normal band-pass filter. However, the higher Q value creates circuit instability, oscillations and makes the circuit very sensitive to the circuit component tolerances. The required Q value, gain, and accuracy required in center frequency determine the practical challenges that we may encounter. While searching for narrow-band filter design documents, I found that only limited literature is available for comparison of various filter performances to enable a designer to select the topology best suited for his applications quickly. In the absence of this literature, one has to resort to simulating all the topologies and end up wasting much time. The selected topology also sometimes does not perform very well practically given many considerations which are generally not documented and comes with experience only. Thus, a seemingly easy electronics design turns a real design challenge in the absence of the right kind of benchmarked comparisons.

This book is written to guide analog electronics designers to understand the performance of popular filters which are typically used to design narrow band-pass filters. The guidelines and performance issues of various filter topologies are explained with a real-world exercise to design an extremely stable narrow band-pass filter to detect a particular very low frequency accurately.

1.2. Design Exercise

Let's consider that the narrow band-pass circuit is required to detect a tone of 577Hz frequency. It requires an extremely narrow band-pass filter which peaks exactly at the desired mid-frequency with an accuracy of ± 6 Hz, This is a low-frequency application, where a 1% error in center frequency means a shift of 6 Hz on either side, thus defeating the design purpose. The bandwidth of 20Hz is required to ensure the power supply harmonics to filter out along with other undesired frequencies. The nearest harmonics of 60Hz power harmonics is 540Hz and 600Hz, and that of 50Hz (US Version) is 550Hz and 600Hz. Maximum 10 Hz bandwidth can be allowed on either side of the mid-frequency to ensure more than 20 dB attenuation to 540Hz and 600Hz frequencies. Above requirement of filtering the harmonics also need a steeper roll-off from 3db points.

A filter may have a desired bandwidth, but it is not guaranteed that the mid-frequency actually occurred at the desired value. In this case, mid-frequency must occur at 577Hz. It should reject the rest of the frequencies outside the bandwidth. The circuit is required to filter out all the rest of the frequencies by 12 dB per decade of attenuation on each side of the mid-frequency. Single supply requirements are important in today's applications as most the electronics circuits, and particularly hand-held units work on batteries and require operation from a single supply.

For any filter design, the Q value determines the performance of the filter. Q value is the ratio of mid-frequency with the bandwidth. Thus, the Q value also determines the maximum allowed bandwidth of the filter. Simply speaking, the desired Q value can be achieved by limiting the bandwidth parameter but real question to investigate during the design is if the circuit peaks exactly at the said mid-frequency. In actual practice, most of the filters exhibit strong Q and gain relationship that changing one will either change another parameter or will shift the center frequency of the filter. Due to component tolerances, the actual mid-frequency could occur anywhere between the two 3dB points in the spectrum. Thus challenge in such designs is if we can keep the mid-frequency stable and insensitive to component tolerance values. For a filter to be highly selective to a given low-frequency, the circuit should have a high Q. However, when working with high Q's one must be very careful with layout and component selection. This is because high Q circuits tend to exhibit instability with a slight component mismatch. They also are more likely to oscillate due to this instability [2]. Also, high Q poses another problem of choosing a right op-amp. With high Q, the op-amp gain-bandwidth product (GBW) can be easily reached, even with a small gain of 20dB. At least 40dB of headroom should be allowed above the center frequency peak [1]. Op-amp slew rate should also be sufficient to allow the waveform at center frequency to swing to the amplitude required. The roll-off requirement needed a minimum 2nd order filter.

The requirements are summarized as:
- Center frequency: 577Hz
- Bandwidth less than 20Hz
- Accuracy & stability of mid-frequency: $\pm$ 1 % ($\pm$ 6 Hz max)
- Roll-off of minimum 12 dB per decade on each side of the center frequency
- Power supply harmonics rejection (540Hz, 550Hz, 600Hz)
- Single supply (+5V) operation
- High overall Q value (Q>25)
- Gain should not be very high (G=5)
- Independent adjustment of Q without affecting center frequency
- Independent adjustment of gain without affecting Q

1.3.   Topologies for Filter Design

A simple survey on the internet would reveal that there are so many topologies for designing the filters [3] [4]. However, the textbooks have very limited literature on all of these topologies. Still, many more complex topologies are being evolved regularly, but some topologies have their established base in popular applications. For filter design, there are many options of topologies which may be potentially used to design the narrow band-pass circuits. Here are some of the most popular options which are categorized based on the count of operational amplifiers (op-amps) required by each of them.

SINGLE OP-AMP TOPOLOGIES
- Sallen-key
- Multiple Feedback (MFB)

- Deliyannis
- Active Twin-T

TWO OP-AMPS TOPOLOGIES
- Fliege

THREE & FOUR OP-AMPS TOPOLOGIES
- State-variable
- Biquad
- Tow-Thomos Biquad
- Akerberg-Mossberg Biquad
- KHN Topology
- Berka-Herpy topology
- Michael-Bhattacharya

The discussion in this book is limited to basic filter topologies. The filter topology derivatives such as Deliyannis (similar to MFB), KHN filter (similar to Biquad in structure) and Tow-Thomos (almost same as Biquad) has not been dealt with here. The Michael-Bhattacharya & Berka-Herpy filters are also not discussed here.

The most desirable situation is, of course, to implement a filter with a single op-amp thus reducing the cost as they may require fewer passive components as well, occupy less space on board and could be tested easily. As a designer, our first preference would be to choose a single op-amp topology. The options available to us are - Sallen-Key, Multiple feedback, Deliyannis, Active Twin-T filters. However, single op-amp topology is associated with many other design considerations. Controlling all the design parameters for single op-amp require intensive calculations and non-standard value of components. As a designer, we would like to have a filter which can work with ordinary components without severely degrading the performance. An additional quad amplifier just cost $0.30 whereas high quality 2% PPS capacitor itself costs around $0.20 each whereas a 1% metal film resistor costs around $0.02 each. So MFB or Twin-T filter requiring two capacitors each requiring high quality 2-3 capacitors and a similar number of resistors would cost more than a multiple amplifiers configuration which are stable enough to work with normal components.

As a design target, complete control over the filter corner/center frequency, the gain of the filter circuit and Q of band-pass filters is required. More control usually means more op-amps, which may be acceptable in designs that are not produced in large volumes, or that may be subject to several changes before the design is finalized (Carter, 2000). While Fliege filter topology is designed with two amplifiers, there are several possible topologies which use three or four amplifiers such as - State-variable, Biquad, Tow-Thomos Biquad, Akerberg-Mossberg Biquad, KHN Topology, Berka-Herpy topology, and Michael-Bhattacharya topology.

In the following sections, the design and performance of following popular filter topologies and compared the relative performance. The book follows the analysis of single op-amp topologies first, leading to two op-amp topology and then filters with multiple op-amps are discussed. The design analysis and performance showed that Biquad filter topology is best suited to meet the stringent requirements outlined in this book to design extremely narrow band-pass filter operating at low frequency. Biquad topology showed very stable performance, however, to meet the stringent design requirements, the design requires further investigations on design optimization to meet roll-off criteria, gain requirements and bandwidth requirements.

Following 8 topologies are discussed and evaluated in subsequent chapters.

1. Sallen-key filter
2. Multiple Feedback (MFB) filter
3. Active Twin-T filter
4. Modified Deliyannis filter
5. Fliege filter
6. Akerberg-Mossberg Biquad filter
7. State-variable filter
8. Biquad/Tow-Thomos Biquad filter

◆ ◆ ◆

CHAPTER 2 – ONE OP-AMP FILTER TOPOLOGIES

2.1   Sallen-Key Topology

At first glance, this topology appears very attractive since it uses only one op-amp a few passive components. This is the simplest and popular topology for which a large amount of literature is available in the textbooks and is easy to integrate quickly [1] [5]. The Sallen-Key topology is one of the most widely known and popular second-order topologies. It is low cost, requiring only a single op-amp and four passive components to accomplish the tuning. There are instances where the Sallen-Key topology is a better choice. As a rule of thumb, the Sallen-Key topology is better if:

1) Gain accuracy is important, and
2) A unity-gain filter is used, and
3) Pole-pair Q is low (e.g., Q < 3)

At unity-gain, the Sallen-Key topology inherently has excellent gain accuracy. This is because the op-amp is used as a unity-gain buffer. The unity-gain Sallen-Key topology also requires only two resistors.

Electronics Design

It employs two RC filters in successions consisting of 3 resistors and 2 capacitors with one op-amp as shown in the figure 1.

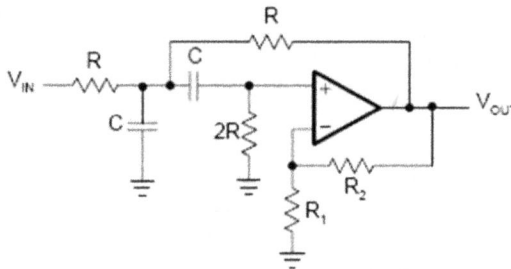

Fig 1: Sallen-key Band-pass Circuit with dual supply

The performance is generally very predictable. A transfer function equation can be derived and gives the values of parameters in terms of components values as under [1]:

$$f_m = \frac{1}{2\pi RC}$$

$$G = 1 + \frac{R2}{R1}$$

$$A_m = \frac{G}{3 - G}$$

$$Q = \frac{1}{3 - G}$$

Where f_m is the mid-frequency, A_m is the gain at mid-frequency, G is the inner feedback gain of the op-amp, Q is the filter quality.

It is obvious from the above that mid-frequency gain is very much dependent on feedback gain and feedback gain is dependent on the value of the feedback resistor. To find the feedback resistor value, either set a fixed value of A_m or fix the value of Q.

To set the frequency of the band-pass, f_m is specified to calculate R as under with an equal value of capacitors C.

$$R = \frac{1}{2\pi f_m C}$$

Because of dependency between Q and A_m, there are two options to solve for R2: either to set the gain at mid-frequency:

$$R_2 = \frac{2A_m - 1}{1 + A_m}$$

Alternatively, to design for a specified Q:

$$R_2 = \frac{(2Q - 1)}{Q}$$

So, the Sallen-Key circuit has the advantage that the quality factor (Q) can be varied via inner gain G without modifying the mid-frequency. A drawback is, however, that Q and A_m cannot be adjusted independently because both are dependent upon the inner gain G. For example, fixing G=2 will give A_m=2, Q= 0.5 and if we want to have high Q lets us say equal to 10, then G=2.9 and A_m will be 29.

Frequency Response/Performance

One can obtain a limited maximum Q value in this circuit, as shown in figure 2. Thus, it is not recommended for applications that need a high Q. For a single op-amp Sallen-Key filter, the Q is typically around 5 or so [6]. Further, this generally works best when the gain is near or a little greater than 1 and Q is less than 3. Changing the gain of a Sallen-Key circuit also changes the filter tuning and the style. It is easiest to implement a Sallen-Key filter as a unity-gain Butterworth. Generally, it gives high gain accuracy with unity-gain [7].

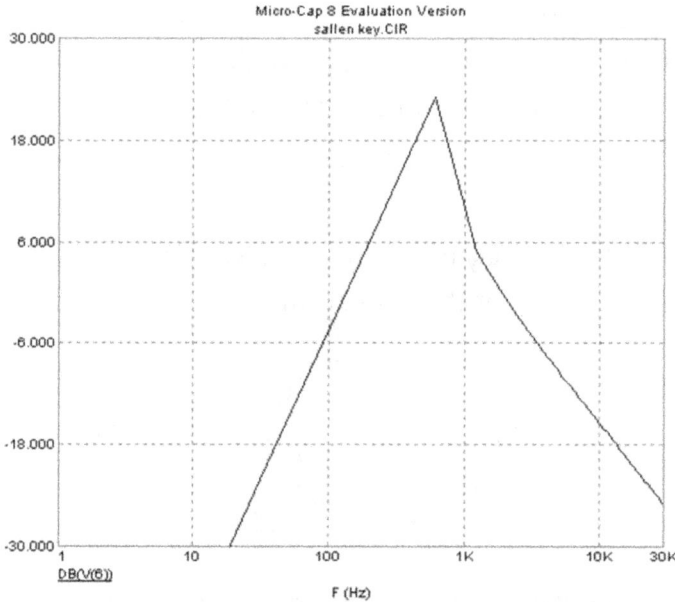

Fig 2: Frequency response of Sallen-Key filter with observed values of
G=2.95, Am=14, Q=3, BW=189Hz, fm=577

Another drawback is that the gain of this circuit is relatively low (−3Q) compared to the minimum required open-loop gain of the amplifier ($90Q^2$). It means that the GBW product of the amplifier must be significantly higher than the maximum cutoff frequency of the filter resulting in a higher performance amplifier than expected to ensure it does not adversely affect the filter response [6].

In the design exercise, there is no suitable tradeoff between A_m and Q. Texas Instruments Single Supply Expert [4], an online design guide for any kind of filter, also does not recommend making high or low Q band-pass filters using Sallen-Key topology.

One more restriction on its use is that it works only in a non-inverting configuration. It poses greater chances of oscillation when G approaches 3, because then A_m becomes infinite and causes the circuit to oscillate. Single supply operation for Sallen-Key BPF would require one buffer op-amp [1]. As it is obvious that it is not suitable for our application as we need a high Q value.

2.2   Multiple Feedback (MFB) Topology

To obtain a slightly higher Q, another simple option is to move to the multiple feedback infinite gain architecture shown in figure 3.

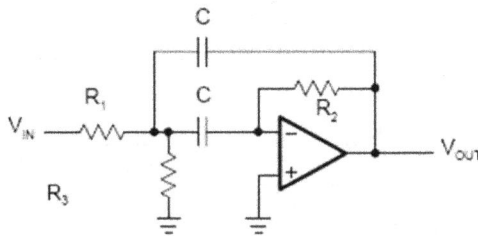

Fig 3: MFB Filter circuit with Dual Supply [1]

The configuration of Sallen-Key and MFB appears similar. Both the topologies are compared briefly in table 1.

Table 1: Comparison of Sallen-Key and MFB topologies [8]

Sallen-Key	Multiple Feedback
Non-inverting	Inverting
Very precise DC-gain of 1	Any gain is dependent on the resistor precision
Less components for gain = 1	Less components for gain > 1 or < 1
Op-amp input capacity must be taken into account	Op-amp input capacity has almost no effect

Resistive load for sources even in high-pass filters	Capacitive loads can become very high for sources in high-pass filters

Electronics Design

This topology again requires a single amplifier and provides for Q's in the range of 25. Using this topology, the gain is ($-2Q^2$) which is still relatively low compared to the amplifiers GBW product ($20Q^2$ at resonance), but not nearly as low as the Sallen-Key approach [6]. Further, it does not require input capacitance compensation. MFB topology is very versatile, low cost, and easy to implement. Unfortunately, calculations are somewhat complex. It gives an easy way of a single supply operation in an inverting configuration [4,7]. The single supply circuit configuration is shown in figure 4.

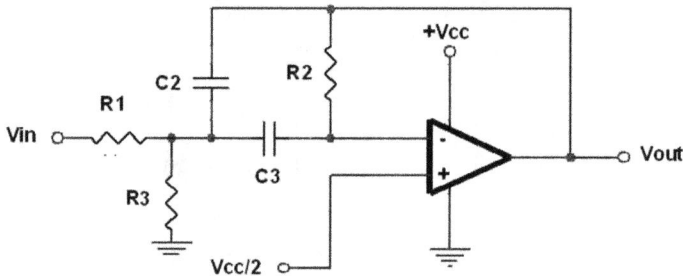

Fig 4: MFB filter with a single supply

The equations for the MFB derived from its transfer function are [1]:

$$f_m = \frac{1}{2\pi C} \sqrt{\frac{R1 + R3}{(R1.R2.R3)}}$$

$$-A_m = \frac{R2}{2R1}$$

$$Q = \pi f_m R2.C$$

$$B = \frac{1}{\pi R2.C}$$

Where B is the bandwidth of the BPF.

The MFB band-pass allows adjusting Q, A_m, and f_m independently. Bandwidth and gain factor do not depend on resistor R3. Therefore, R3 can be used to modify the mid-frequency without affecting bandwidth, B, or gain, A_m [1]. However, when we simulate this circuit, we will observe that the change in R3 would cause a change in gain also.

For low values of Q, the filter can work without R3. However, Q then depends on A_m via:

$$-A_m = 2Q^2$$

This topology without R3 is called Deliyannis [3] and is useful for low Q values only. However, Deliyannis has an extra gain resistor set too.

The components value can be fixed as under if we fix f_m, gain and Q values:

$$R1 = \frac{R2}{2A_m}$$

$$R2 = \frac{1}{\pi BC} = \frac{Q}{\pi f_m C}$$

$$R3 = \frac{R2A_m}{2(Q^2 + A_m)}$$

Frequency Response/Performance

This topology inverts the signal and also note that the gain and Q value are inextricably related as with the Sallen-Key [1]. It is used when high gain and high Q values are needed. Since we need a high Q (and if possible high gain too), MFB appeared to be a good choice. The figure 5 shows the frequency response of the MFB filter, the sharp peak, and steeper roll-off are the other parameters about which we should be concerned.

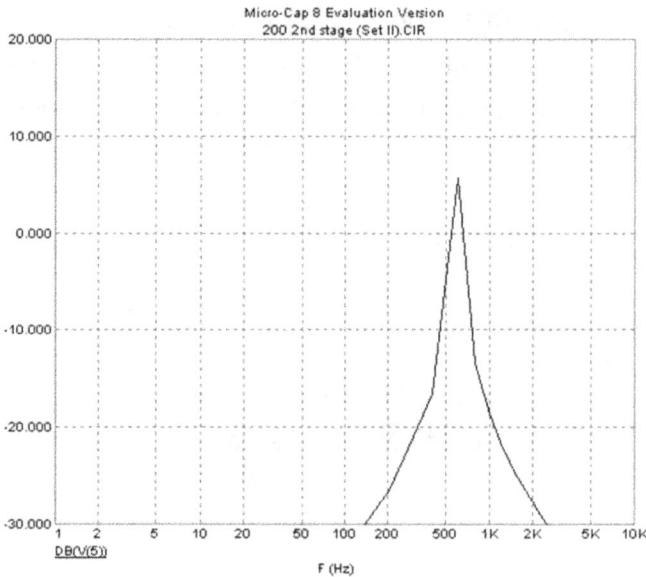

Fig 5: Frequency response of MFB filter with G=2, Q=15, BW=35, fm=577, 2nd order Narrow BPF

Both the Sallen-Key and the multiple feedback architectures are fairly sensitive to external component variation [4]. MFB particularly is very sensitive to variation in R3 resistor. Higher the Q value the smaller will be the value for the resistor. At very high Q values, the value of this resistor can be in few hundred ohms. Very precise resistors and capacitors are needed to make a narrow band-pass filter with MFB topology. Further, the overall response of the MFB filter over the tolerances of the capacitors and resistors is also very important. The Monte Carlo simulation results for the MFB filter is shown in figure 6.

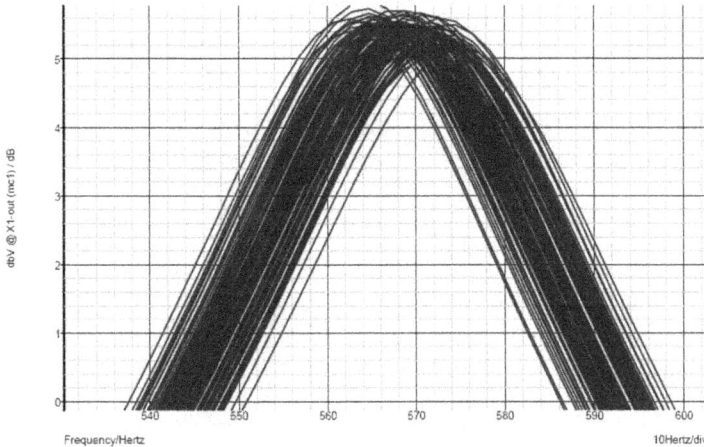

Fig 6: Monte Carlo Simulation of MFB Filter over 1% resistor tolerances and 2% capacitor tolerances

This simulation is showing a variation of mid-frequency from 561Hz to 577Hz with 1% resistors and 2% capacitors. Stability factor is a very important aspect here when working on high Q values. MFB scores better than Sallen-Key topology on many factors and parameters.

At this point, it is worth mentioning that Texas Instruments' Analog Filter expert site [4] does not recommend this topology for low as well as high Q band-pass filters. Experimental setup of the MFB prototype circuit indicates that MFB was very sensitive to the tolerances of the external components, especially attenuator resistor R3.

2.3   Active Twin-T Topology

This topology is also very attractive but is known mostly for a notch filter configuration. However, this topology also works very well in a narrow band-pass mode. The beauty of this topology is that it can specifically be used to achieve narrow peaks. Most of the single tone detector circuit employs Twin-T topology, so this would be a topology worth making a study in a little more detail in order to check if this could be a suitable topology for extremely narrow band-pass filters.

This topology is based on two passive (RC) Twin-T circuits as shown in figure 7, each of which uses three resistors and three capacitors [7].

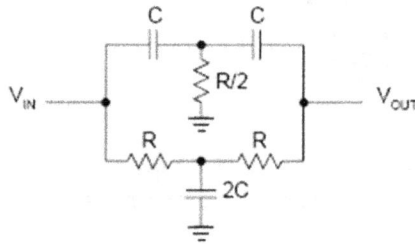

Fig 7: Two R-C T circuits to make Twin-T circuit

The most important thing is that the matching these six passive components are critical; fortunately, it is also easy. Components from the same batch are likely to have very similar characteristics. The entire network can be constructed from a single value of resistance and a single value of capacitance, running them in parallel to create R/2 and 2C in the Twin-T schematics [7]. The sharpest response will come when the components are matched (by using components from the same batch and by creating R/2 and 2C by paralleling 2 of the values used for R and C, respectively).

Electronics Design

In total, it requires 6 passive components and minimum one op-amp. The Twin-T topology can be made with one as well as two op-amps [3]. One op-amp configuration is shown in figure 8, and two op-amps configuration is shown in figure 9.

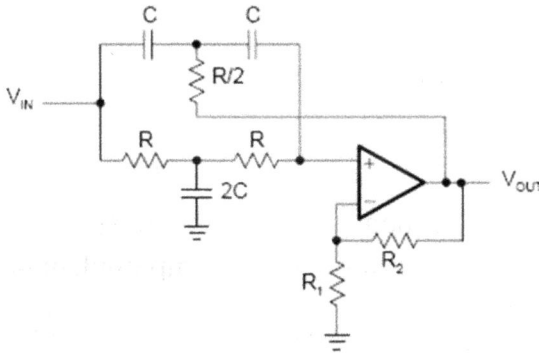

Fig 8: Twin-T band-pass filter with dual supply

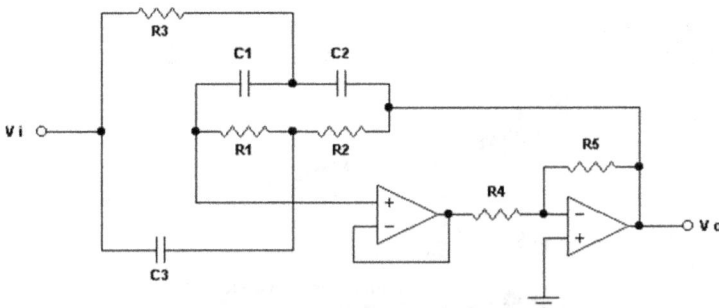

Fig 9: Two op-amp implementation of Twin-T BPF with dual supply

The Twin-T circuit has the advantage that the quality factor Q can be varied via the inner gain G without modifying the mid-frequency f_m. However, Q and A_m cannot be adjusted independently [1].

To set the mid-frequency of the band-pass, specify f_m and C, and then solve for R:

$$R = \frac{1}{2\pi f_m C}$$

Because of the dependency between Q and A_m, there are two options to solve for R2 - either to set the gain at mid-frequency:

$$R2 = (A_m - 1)R1$$

Alternatively, design for a specific Q:

$$R2 = R1(1 - \frac{1}{2Q})$$

As mentioned earlier non-inverting and inverting both versions are available. Further, there is also two op-amp implementation found in the literature [3,7].

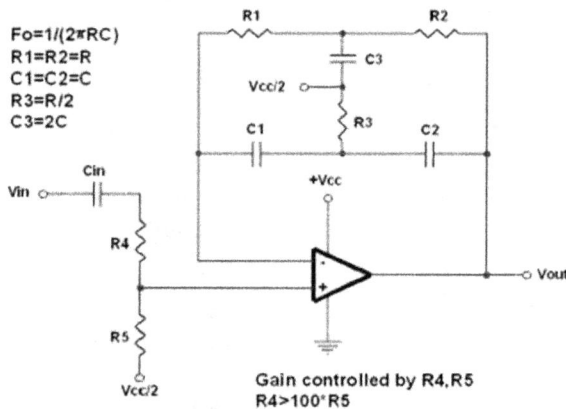

Fig 10: Twin-T BPF with a single supply

For single supply operation, the circuit configuration is changed a little (refer to figure 10) at input side [4, 7] where R4 and R5 control the gain and R7 is the tuning resistance for adjusting Q value. However, Q cannot be changed freely without affecting other parameters. It has been observed that Q is hard to control in this circuit [4, 7]. Absolute levels of amplitude will be hard to obtain [4, 7].

It was mentioned just in the beginning of this section that matching is required for the perfect functioning of Twin-T filter. However, in practice, this high Q band-pass circuit made with Twin-T would oscillate and become unstable if the components are matched too closely [7]. In order to have better control of Q, the components need to be mismatched slightly and best to de-tune it slightly. By selecting the resistor to virtual ground to be one E-96 1% resistor value off, for instance. It may be noted that mismatching would affect the gain also [4].

As per Texas Instruments' Single Supply Analog Filter online guide, theoretically, this circuit is considered to be most suitable for use as a high sensitivity single tone detector [4]. Incidentally, the application intended by us has been the same. Practically we observe that this circuit needs much care in component selection while designing. High precision components add to the cost of the circuit.

Frequency Response/Performance

Before finalizing the circuit, we want to evaluate the other topologies as well. In the absence of any other stable topology, Twin-T could have been the strong contender for the application along with MFB. The frequency response of Twin-T is surely good enough (figure 11), but the roll-off is not as steeper as MFB. In that regard, MFB scores better than Twin-T in most of the aspects. In that case, a tradeoff could have been made between MFB and Twin-T w.r.t to their relative performance.

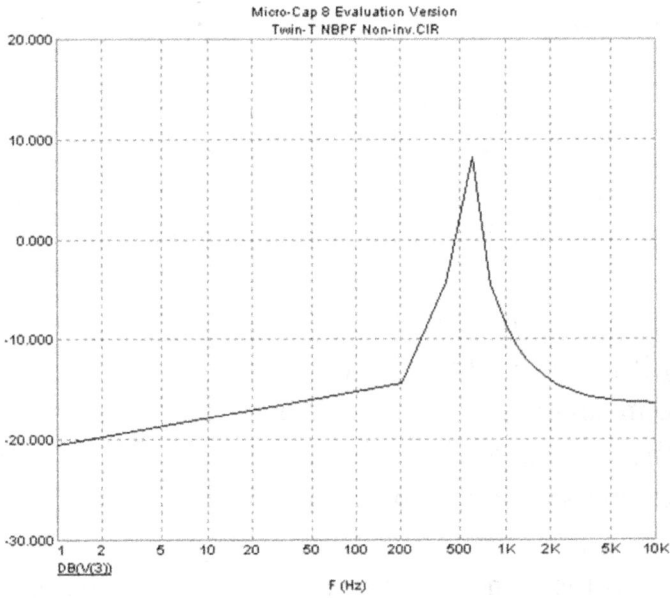

Fig 11: Frequency response of Twin-T BPF with G=10, Q=10, BW=50, fm=577

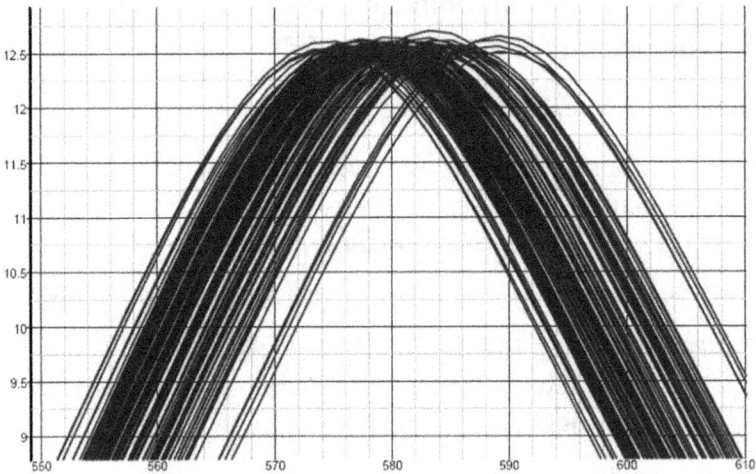

Fig 12: Monte Carlo Simulation of Twin-T Filter over 1% resistor and 2% capacitor tolerances

A Monte Carlo analysis as shown in figure 12 also shows that there are bigger variations in center frequency concerning tolerance. The center frequency with the same tolerance as used for MFB, however, center frequency can be seen varying from 572 Hz to 588Hz, which indicate that Twin-T is more sensitive to component tolerances.

However, still, many more topologies are available which need to be evaluated.

2.4   Modified Deliyannis Topology

The Deliyannis filter shown in figure 13 is just the MFB modified filter with an attenuator resistor R3 missing [3]. Ideally, that combination of MFB is used when low Q value is needed, as it was pointed out in the discussion of MFB. However, with some modification in basic Deliyannis, a high Q value is possible to achieve [9].

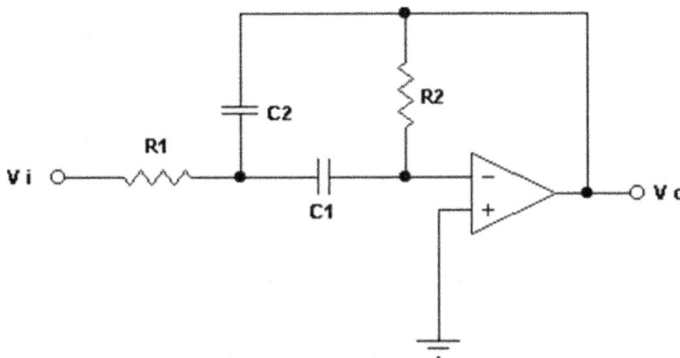

Fig 13: Basic Dalianis Band-pass filter with dual supplies

Electronics Design

The feedback resistor in MFB is split into 2 and attenuator resistor is connected as it is as shown in figure 14.

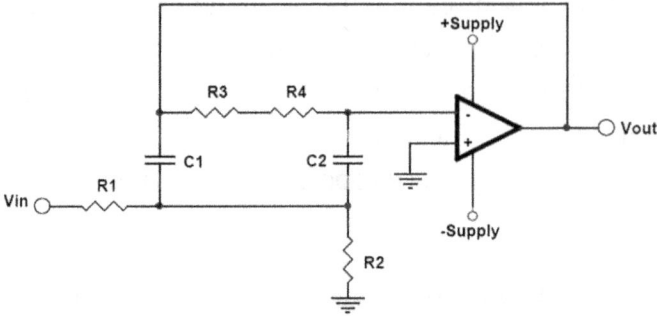

Fig 14: Modified Dalianis Band-pass filter with dual supplies and feedback resistor split into two resistors

The circuit works on a single supply and its configuration with a single supply would be as shown in figure 15.

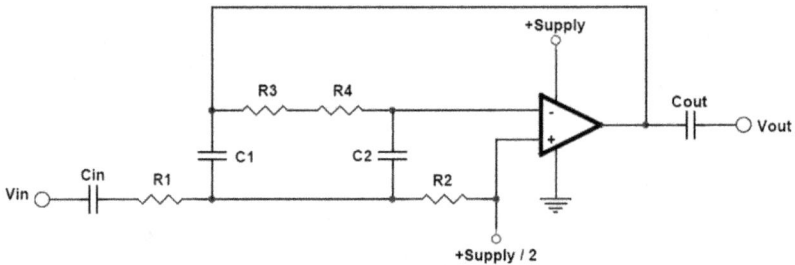

Fig 15: Dalianis Band-pass filter with a single supply

For the simplified circuit, we choose C1=C2 and R1=R4= R in such a way that

$$R = \frac{1}{2\pi f_m C}$$

The gain and Q of the circuit are interlocked and controlled by R1 and R4 through the equation [9]:

$$G = \frac{R3 + R4}{2R1} = Q$$

The resistor R3 depends upon the gain value and can be selected using the equation:

$$R3 = (2A_m - 1)R1$$

Because gain and Q are linked together, gain resistors R1 and R3 can be used as a voltage divider to reduce the input level and compensate for this effect as shown in figure 15. When gain and Q approach one, R1 can be shorted and R2 can be opened [12].

If R3 is doubled, R2 must be halved and vice versa. If one is tripled, the other must be one third, etc. R2 and R3 must always be related in this way. Otherwise, the center frequency and other circuit characteristics are changed.

$$n. R3 = \left(\frac{1}{n}\right) R2$$

If R1 = R2 = R3 = R4, then Q and gain are both equal to one.

Frequency Response/Performance

This filter acts as BPF as shown in its frequency response diagram figure 16, with a very sharp peak, the lower roll-off is not as steep as MFB, but the gain is much higher than MFB.

A Monte Carlo analysis in figure 17 indicates a variation in center frequency from 570Hz to 588Hz which is comparable to MFB and better than Twin-T.

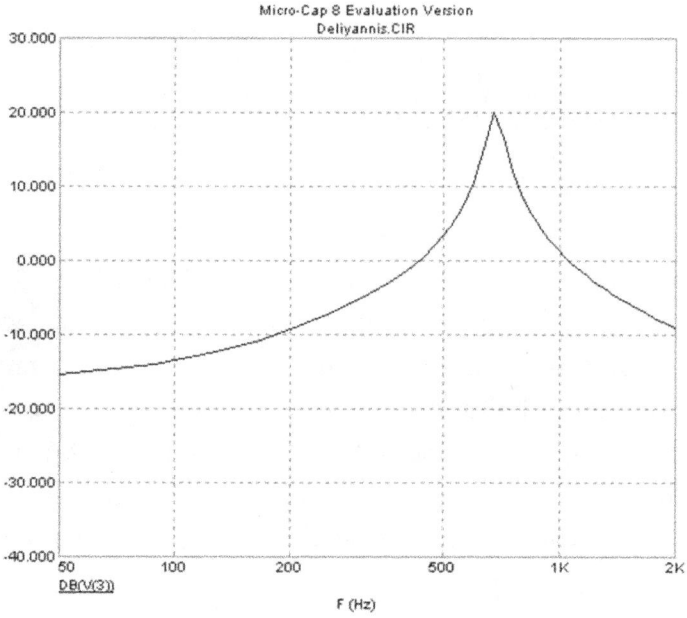

Fig 16: Frequency response of Deliyannis BPF with G=10, Q=10, BW=50, f_m=577

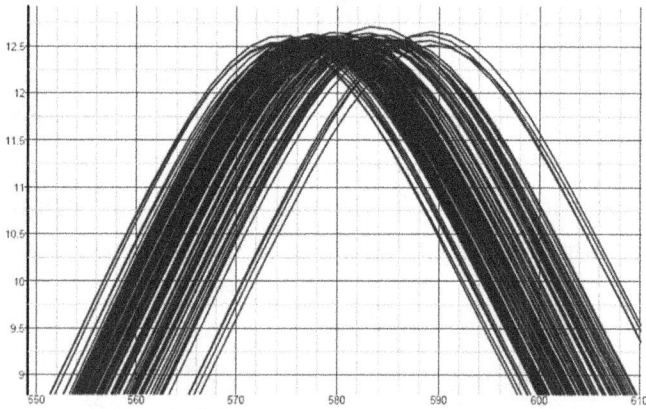

Fig 17: Monte Carlo Simulation of Deliyanis filter over 1% resistor and 2% capacitor tolerances

However, the only obstacle has been the gain and Q being same [9]. In order to keep the Q high, the gain has to be kept high. The Dalianis is supposed to be better in terms of the variation due to component tolerances. However, the circuit performance is not as good as MFB and not very much suitable for our application. MFB still score better than Deliyannis topology.

◆ ◆ ◆

CHAPTER 3 – TWO OP-AMPS FILTER TOPOLOGIES

3.1 Fliege Topology

A Fliege Filter is a two-pole filter topology [4]. It is available in low-pass, high-pass, band-pass, and notch versions. It is a two-op-amp topology as shown in figure 18, as and is more expensive than one op-amp topologies [7].

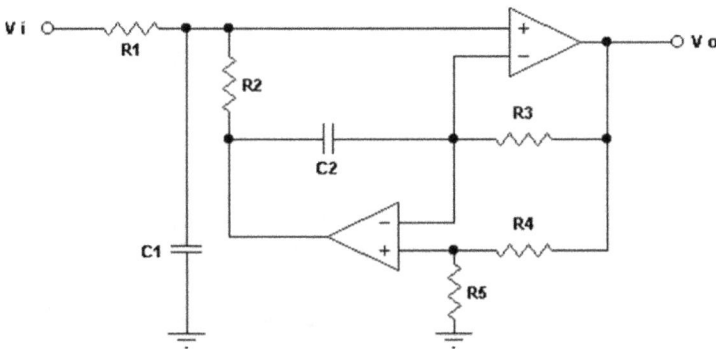

Fig 18: Fliege BPF with dual supply

There is a good control over the tuning and the Q and style of filter. The low-pass, high-pass, and band-pass versions have a fixed gain of

2, while the notch version has unity gain. The Q of the notch version cannot be adjusted.

Electronics Design

The Fliege topology has the lowest component count out of the 2 op-amp topologies and offers good controllability over frequency and type (Butterworth, Chebyshev, and Bessel).

The Fliege topology is well suited to operation from a single supply [4]. Current feedback amplifiers cannot be used, because a capacitor is connected from the op-amp output to inverting input, see figure 19.

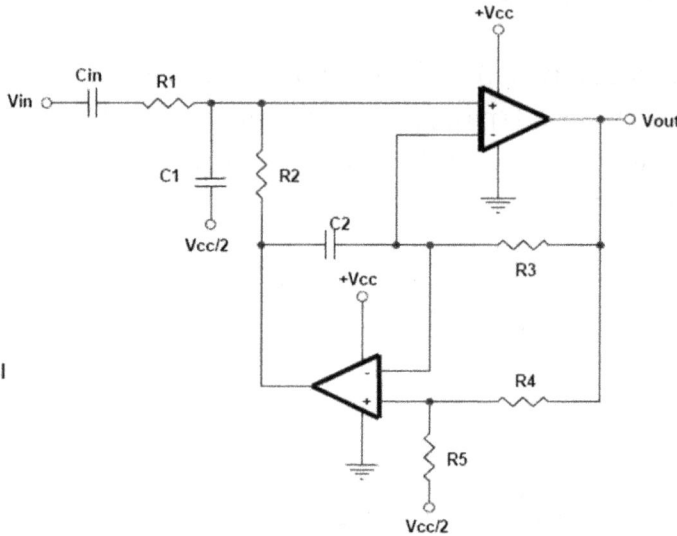

Fig 19: Fliege BPF with a single supply

For simplified circuit we choose, R2 = R3 = R and R4 =R5 so that

$$R = \frac{1}{2\pi f_m C}$$

The circuit offers an excellent option of controlling the Q value by a single resistor R1. For high Q value of R1 is to be chosen high [7]. It will not affect the gain or the mid-frequency. The following equation gives Q value:

$$Q = \frac{R1}{R}$$

Frequency Response/Performance

The circuit performance in terms of Q control matched quite near to our requirements. The gain of this topology is fixed at 2, which could be increased by using the additional op-amp at the output. In that case, it would have become a three-op-amp topology. The initial results of Fliege were very encouraging. A very sharp peak could be received as seen from the figure 20 which is quite similar to the MFB filter.

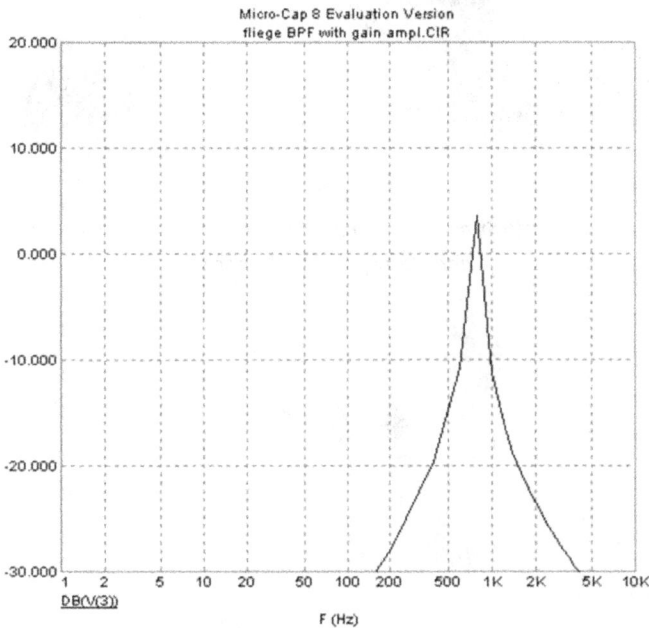

Fig 20: Frequency response of Fliege Filter, Gain =2, Q=5, f_m. 577 HZ

The big trouble with this circuit turns out to be its sensitivity to component tolerances. Even the small tolerances shifted the mid-frequency by a large amount.

A Monte Carlo simulation analysis figure 21 showed even if we use 1% resistors and 2% capacitors, the mid-frequency may vary from 573 to 590 Hz, which is a little worse than MFB. The temperature sensitivity of the circuit also indicated that the circuit is not stable at mid-frequency even if it is giving the sharp peak. In terms of stability, this topology was not considered for implementation.

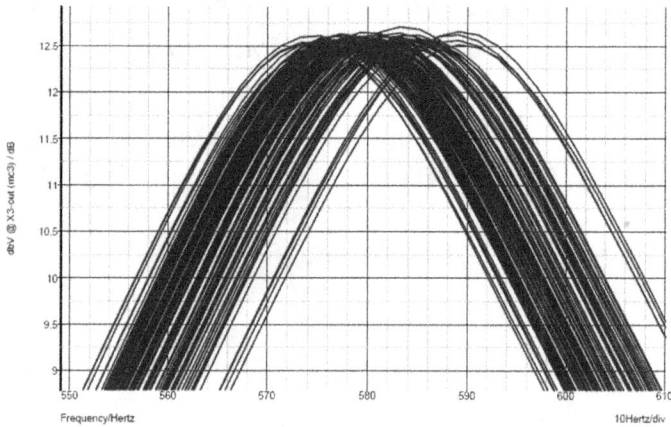

Fig 21: Monte Carlo Simulation of Fliege Filter over 1% resistor tolerances and 2% capacitor tolerances

◆ ◆ ◆

CHAPTER 4 – SUMMARY OF PERFORMANCE OF ONE AND TWO OP–AMPS FILTER TOPOLOGIES

4.1 Performance Comparison

Based on a detailed analysis of one and two op-amp configurations, the summary of evaluation is as follow:

Sallen-Key

This topology appears very attractive since it uses only one amplifier and a few passive components. Mid-frequency gain is very much dependent on feedback gain, and feedback gain is dependent on the value of the feedback resistor. So, the Sallen-Key circuit has the advantage that the quality factor Q can be varied via inner gain G without modifying the mid-frequency [5]. A drawback is, however, that Q and A_m cannot be adjusted independently because both are dependent upon the inner gain G. Another drawback is that the gain of this circuit is relatively low (-3Q) compared to the minimum required open-loop gain of the amplifier ($90Q^2$). This means that the GBW product of the amplifier must be significantly higher than the maximum cutoff frequency of the filter resulting in requiring a higher performance amplifier than expected to ensure it does not adversely affect the filter response [12]. Due to these drawbacks; it could not fit into our requirements.

MFB

MFB topology is very versatile, low cost, and easy to implement and allows to adjust Q, A_m, and f_m independently. Bandwidth and gain factor do not depend on the attenuator resistor. Therefore, this resistor can be used to modify the mid-frequency without affecting bandwidth, B or gain A_m. However, change in R3 would cause a change in gain as well. The gain and Q value are very much related to each other as is the case with Sallen-Key [1]. It is used when high gain and high Q values are needed. Since we needed a high Q (and if possible high gain too), MFB appeared to be a good choice for extremely narrow band-pass filter design. The frequency response of the MFB filter exhibits the sharp peak and steeper roll-off. The Monte Carlo simulation results for the MFB filter shows a variation of mid-frequency from 570 to 580 with 1% resistors and 2% capacitors. MFB particularly is very sensitive to variation in attenuation resistor, but not to other component variations and very precise resistors, and capacitors are needed to make a narrow band-pass filter with MFB topology (Elliot, 2000). Thus, the variation of mid-frequency due to tolerances of the components made it unfit against stiff requirements.

Modified Deliyaanis

Deliyaanis filter is just the MFB modified filter with an attenuator resistor missing [3]. This filter acts as BPF as with a very sharp peak, the lower roll-off is not as steep as MFB, but the gain is much higher than MFB. A Monte Carlo analysis indicates a variation in center frequency from 574Hz to 588 Hz which is comparable to MFB and better than Twin-T. The only obstacle would have been the gain and Q being same. In order to keep the Q high, the gain has to be kept high. The Daliyaanis is supposed to be better in terms of the variation due to component tolerances. However, the circuit performance is not as good as MFB and not very much suitable for our application.

Active Twin-T

The Twin-T circuit has the advantage that the quality factor Q can be varied via the inner gain G without modifying the mid-frequency f_m. However, Q and A_m cannot be adjusted independently [1]. The most important thing is that the matching these six passive components is critical. However, in practice, this high Q band-pass circuit made with Twin-T would oscillate and become unstable if the components are matched too closely. The frequency response is surely good enough, but the roll-off is not as steeper as MFB. A Monte Carlo analysis shows that there are bigger variations in center frequency varying from 572 Hz to 588Hz, which indicate that Twin-T is more sensitive to component tolerances and hence unfit for our application.

Fliege

This is the lowest component count for two op-amp topologies. This offers an excellent option of controlling the Q value by a single resistor R. A very sharp peak could be received which is quite similar to MFB filter. The big trouble with this circuit turned out to be its sensitivity to component tolerances. Even the small tolerances shifted the mid-frequency by a large amount. A Monte Carlo simulation analysis showed even if we use 1% resistors and 2% capacitors, the mid-frequency may vary from 573 to 590 Hz, which is a little worse than MFB. The temperature sensitivity of the circuit also indicated that the circuit is not stable at mid-frequency even if it is giving the sharp peak. The stability is the biggest concerns that this topology is not suitable for our application.

The summary of observations on 1- or 2-op-amp topologies filters as seen from literature, simulation, calculations, and experimentation is summarized in the following table 2. The properties of interest to a designer while designing a band-pass filter could be independent control of Q, f_m and A_m, the sharpness of peak, stability, and accuracy of the mid-frequency, sensitivity to component variations, types of components and op-amps required, single supply operation and oscillations.

Table 2: Comparison of various properties of topologies for suitability for extremely narrow and pass filter against specified requirements

Property	Sallen-key	MFB	Deliyaanis	Active Twin-T	Fliege
Number of op-amps	1	1	1	1or 2	2
Q value obtainable	Lowest (Q =3 to 5)	Medium (Q>25)	Low to high (Q>0.5 to Q =100)	Highest (Q>100)	High
Dependence of Q and A_m (mid-frequency gain)	Strong A_m=3Q	Strong A_m=2Q^2	Yes (interlocked and equal) A_m=Q	Yes	No
Flexibility to change Q via inner gain	Yes Q= 1/ (3-G)	No	Yes Q=G	Yes	No
Chances of oscillations at high Q	Low	Medium	Low	Highest	Medium
Gain value obtainable	High (1 to X)	High (1 to 10)	Higher (1 to Q)	Highest (>10)	Low (fixed 2)
Gain accuracy	High at unity-gain	Low (depends upon resistor precision)	Low	Low	High
Provision of increasing inner gain in the circuit	Yes	Yes	Yes	Yes	No
Possible narrow Bandwidth	>100Hz	<30 Hz	<30 Hz	<20 Hz	<10 Hz

Independent control of mid-frequency f_m without affecting BW or Q	- - - -	Yes (via attenuator resistor)	- - - -	- - - -	Yes (via input resistor)
the sharpness of Mid-frequency curve peak	no	yes	Yes	Yes	Yes
Errors in mid-frequency due to 2% component tolerances	- - -	1.3 %	- - -	- - - -	1.7%
Lower and upper roll-off	No	Highest on both sides	High on one side and low on other	Low	High on both sides
single supply operation	No	yes	Yes	Yes	Yes
Required Gain-Bandwidth product of amplifier	High ($90Q^2$)	Low ($20Q^2$)	-	-	-
Input Capacitance Compensation required	Yes	No	No	No	Yes
Number of passive components required	7+1	5 +3	6+1	6+3	8
Sensitivity to external component variation	Low (least at G=1)	High (mainly to one resistor)	Medium	Highest	High
Need for precision components	No (Low)	Yes (high)	Yes (Medium)	Yes (Highest)	Yes (Highest)

Still above table can only be considered as a guideline and is nevertheless comprehensive given the tradeoffs involved between various interlinked parameters. Further, the relative importance of various parameters to be traded-off depends on the nature and type of the application, stringent requirements thereon and criticality of the results sought from the filter.

4.2   Tradeoffs in Selection

While selecting the right topology out of these topologies, we would intend that the implementation of the filter should be such that we would have complete control over:
· The filter corner/center frequency
· The gain of the filter circuit
· The Q of band-pass filters

Unfortunately, complete control over the filter is seldom possible with a single op-amp. Even if control was possible, it frequently involves complex interactions between passive components, and this means complex mathematical calculations that intimidate many designers. If the designer needs to implement a filter with as fewer components as possible, there will be no choice but to resort to traditional filter design techniques and perform the necessary calculations.

The most desirable situation is, of course, to implement a filter with a single op-amp thus reducing the cost, but at the same time, we would like to have a filter which can work with ordinary components without severely degrading the performance. An additional quad op-amp just cost $0.30 whereas high quality 2% PPS capacitor itself costs around $0.20 each whereas a 1% metal film resistor costs around $0.02 each. So MFB or Twin-T filter requiring two capacitors each requiring high quality 2-3 capacitors and a similar number of resistors would cost more than a multiple op-amps configuration which is stable enough to work with normal components

4.3   Cascading Multiple Stages of One or Two Op-amps

We find that MFB could be a good topology to use with right kinds of components. The Sallen-Key and MFB architectures also have tradeoffs associated with them. The simplifications that can be used when designing the Sallen-Key provide for an easier selection of circuit components, and at unity-gain, it has no gain sensitivity to component variations. The MFB shows less overall sensitivity to component variations and has a superior high-frequency performance.

Instead of resorting to multiple op-amps configurations, if a designer would like to stick to single op-amp topologies (reason being the simplicity of architect and easy calculations), then he could cascade two stages of similar topology together with little-relaxed design restrictions. The possible solutions with cascading of one op-amp topologies would depend upon the fact whether the gain is independent of Q or not. In case gain and Q are independent and the trouble is to counter instability due to the high Q of a single stage, then we could go for a solution which involves:

- Use two stages cascaded together; this will require two op-amps
- Use the low value of Q for each stage; cascading will give a steeper roll-off and hence better Q
- Provide limited gain to each stage and if needed add additional gain amplifier stage

Such a solution is depicted in figure 22.

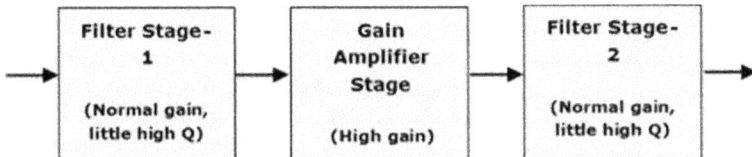

Filter Stage-1	Gain Amplifier Stage	Filter Stage-2
(Normal gain, little high Q)	(High gain)	(Normal gain, little high Q)

Fig 22: Cascading of Filter stages

Another scenario could be that Q and G are interrelated so that only high gain can give high Q values. Then we can think of keeping the Q and gain both at normal values and attenuating the gain after cascading. Whatever the case may be, from a performance point of view, the cascading gives much better results in terms of higher gain (20Db), high Q value, narrower bandwidth and sharper roll-off. Since the Q curve become sharper due to an extra roll-off and gain control remains independent, our intentions at design goals would be fulfilled. The above solutions are definitely going to make the number of components little more than whatever is required for a 2 or 3 op-amp topology, but it may be worth doing it.

Going one step further, two different kinds of topologies each working at relaxed Q and gain restrictions could be cascaded to take advantage of the multiple topologies in one design. For example, a high-performance filter design may be a combination of MFB and Sallen-Key sections. This flexibility can be quickly leveraged by some popular filter design software such as FilterPro to define the component values of the same design using both circuit types and then use some sections from one topology design with some from the other topology design to build the filter. Thus, cascading of same or different topologies can give better results. Such a tradeoff could pay off in terms of stable circuit parameters, a nominal number of passive components, ease of design modifications in later phases and sticking to the basics of the filter design science.

◆ ◆ ◆

CHAPTER 5 – THREE AND FOUR OP–AMPS FILTER TOPOLOGIES

5.1   Multiple Op–amps Topologies

As a designer, if one has made up the mind for cascading, it would be worth analyzing two op-amp and three op-amp topologies too. Even the two op-amp implementation like Fliege topology gives very good results comparable to MFB or Active Twin-T in terms of peaking, sharpness and the number of passive components being used [1,4]. On the other hand, most of the Biquad based multiple op-amps topologies too, like State-variable, Akerberg-Mossberg and Biquad etc, also score much better on stability, immunity to component tolerances and high Q values independent of gain [6, 7].

More control usually means more op-amps, which may be acceptable in designs that will not be produced in large volumes, or that may be subject to several changes before the design is finalized [7]. When cascading is the necessity in order to produce stability in the mid-frequency performance, immunity to component tolerances and elimination of circuit oscillations, the multiple op-amp topologies such as State-Variable, Akerberg-Mossberg, Biquad, Tow-

Thomos designs are worth considering. These designs outweigh single op-amp topologies on achieved performance vs. spent cost scale.

Akerberg-Mossberg topology shall be briefed at the beginning followed by detailed observations on only two topologies namely: State-Variable and Biquad, have been given here. We found that Tow-Thomos was a simple derivative of Biquad and Akerberg was a variation of Tow-Thomos topology. However, State-variable and Biquad were chosen for experimentation.

5.2   Akerberg-Mossberg Topology

This is the easiest of the three-op-amp topologies to use and one of three topologies that offer complete and independent control over gain, frequency, and type (Butterworth, Chebyshev, and Bessel).

Electronics Design

An Akerberg-Mossberg filter is a two-pole filter topology which is implemented by using three op-amps as shown in figure 23.

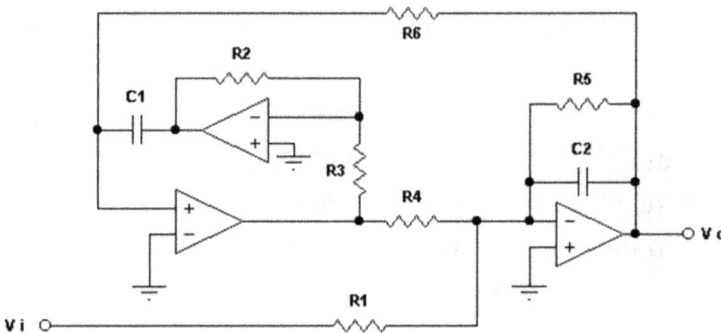

Fig 23: Akerberg-Mossberg Filter using Dual supply

It is available in a low-pass, high-pass, band-pass, and notch versions [3]. It is easy to change the gain, style of low-pass and high-pass filter, and the Q of band-pass and notch filters [4,7]. Gain can be unity or can be fixed any value. Signal input is at different places for the different versions, but the output is always taken from the same point.

The Akerberg-Mossberg topology is suited to operation from a single supply [4,7]. Current feedback amplifiers cannot be used because a capacitor is connected from the op-amp output to inverting input as shown in figure 24. The Akerberg-Mossberg topology can be used with fully differential amplifiers, with the additional advantage that the number of op-amps required is reduced from 3 to 2 [4].

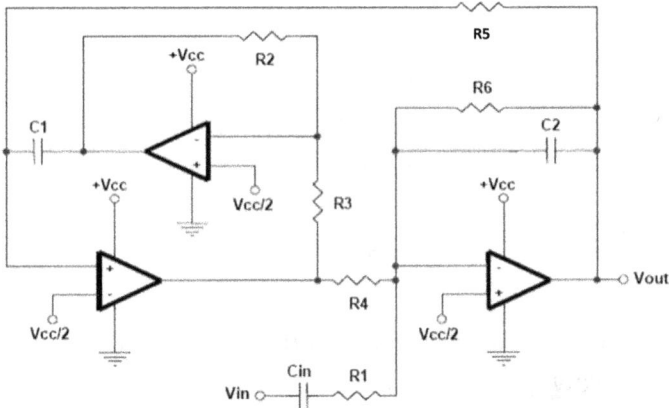

Fig 24: Akerberg-Mossberg Filter using single supply

For the simplified circuit, we choose R2=R3=R4=R5= R and C1= C2= C in such a way that:

$$R = \frac{1}{2\pi f_m C}$$

The mid-frequency gain is given by:

$$A_m = \frac{R6}{R1}$$

The R1 and R6 also control the Q values. High values of R1 and R6 means high Q values. The beauty of this topology is that despite the use of 3 op-amps, it just needs 9 passive components.

Akerberg-Mossberg Topology is a modification of Tow-Thomos topology [3] which itself is a derivative of Biquad filter [10]. This may be clearer when the circuit diagram of Akerberg-Mossberg filter is drawn in Biquad fashion as shown in figure 25.

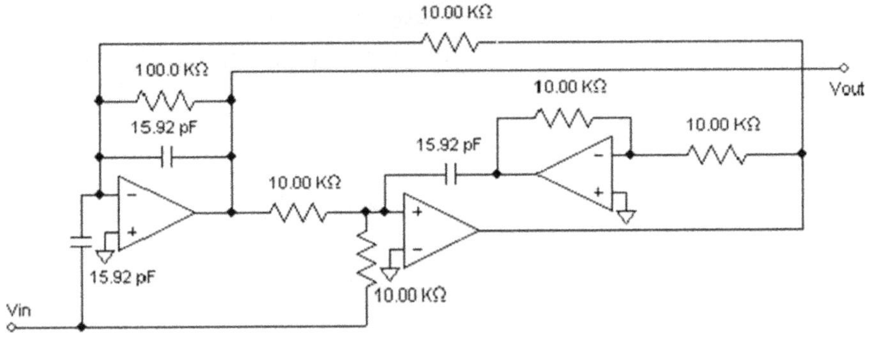

Fig 25: Akerberg-Mossberg BPF drawn in a Biquad fashion

The mathematical transfer functions and associated literature on Akerberg and Tow-Thomos is very limited, of course, it is easy to implement these stages and get the desired mid-frequency just by choosing one value of all the resistors and one value of all the capacitors. This prompts us to look for Biquad type of topology for our application.

5.3   State-variable Topology

The state-variable is a three to four op-amp topology. The fourth op-amp is only required for notch filters. It is also very easy to tune, and it is easy to change the style of low-pass and high-pass, and easy to change the Q of the band-pass and notch. Unfortunately, it is not as nice a topology as Akerberg-Mossberg [7]. The same resistor is used for gain and style of filter/Q, limiting control of the filter. There is probably not much reason to use this topology unless the application requires simultaneous low-pass, high-pass, band-pass, and notch outputs. It uses more number of op-amps, but integrate low-pass, high-pass and band-pass filters in one and any transfer functions can be realized by combining the outputs.

Electronics Design

Figure 26 shows the basic architecture of the 3-amplifier State-variable Biquad. It comprises a summing node followed by two integrators. This architecture is quite versatile in that it gives a high-pass, band-pass and low-pass output, but it also allows independent control of f_m, and the Q [6].

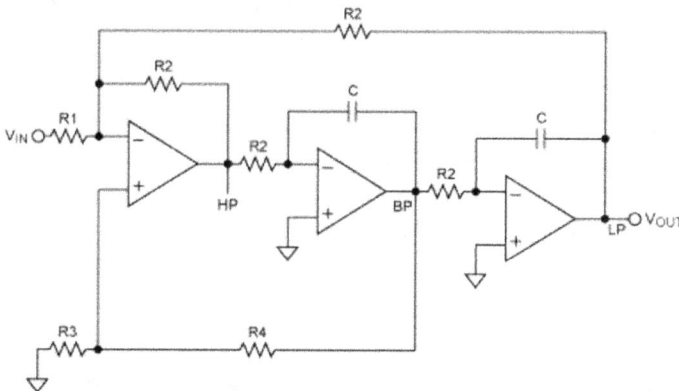

Fig 26: Three amplifiers State-variable Biquad with dual supply

The configuration can work very well with a single supply [4]. The single supply circuit is shown in figure 27.

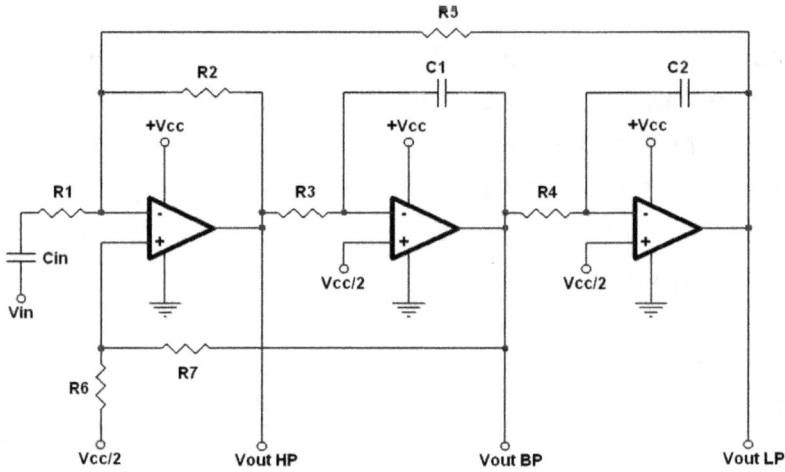

Fig 27: Three amplifiers State-variable Biquad with a single supply

The circuit can be integrated with a single value of the capacitor and resistors by choosing R1=R2=R3=R4=R5=R6=R and C1=C2=C in such a way that:

$$R = \frac{1}{2\pi f_m C}$$

R7 controls the gain since mid-frequency is given by the equation [4]:

$$-A_m = \frac{R7}{R}$$

For better performance value of R7 is chosen in such a manner that R7= 3.Q.R, so roughly G= 3.Q. Higher values of R7 ensure a high Q while lower values of R7 will give low Q value. R7 have another important impact on the type of HP and LP filter. If R7> R/2, the LP and HP filters type turn out to be Chebyshev [7]. For steeper roll-off in BP filters, R7 is chosen to be greater than R/2. It is worth mentioning that BP output is coming due to the superimposition of HP and LP filters. Thus the style of these LP and HP filters will control the steepness in the roll-off in BP filter too [11].

Frequency Response/Performance

The response due to the superimposition of LP and HP filter is shown in figure 28.

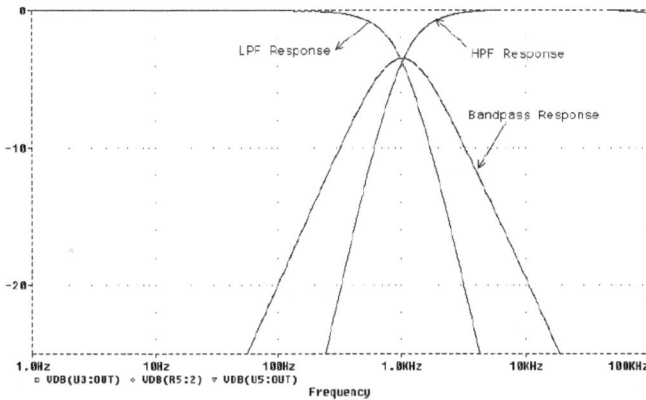

Fig 28: Band-pass filter response of Biquad due to superimposition of LP and HP response [11]

The frequency response, figure 29 of State-variable does not seem to be encouraging. However, one major point observed is the flatness of the Band-pass, which indicate that that the filter performance is going to be very stable in band-pass mode.

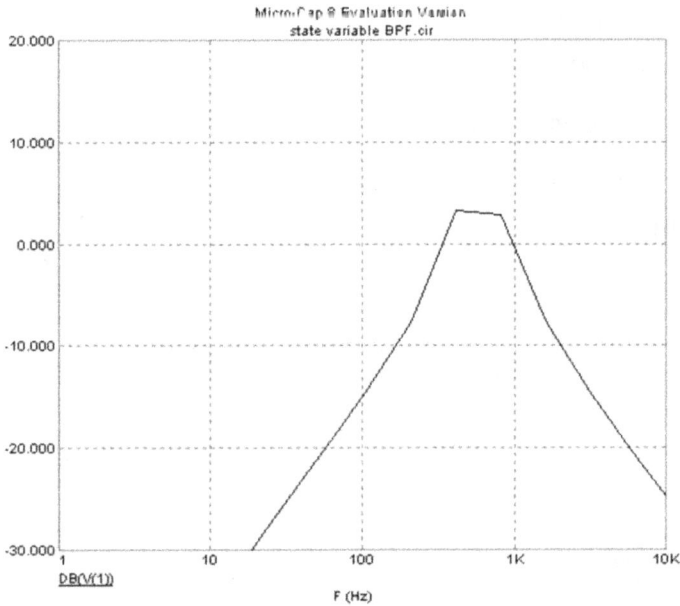

Fig 29: State-variable BPF response without independent Q control. The filter
shows a wide and almost flat pass-band with a steep roll-off on the sides.

Modified State-variable Filter

It may be noted that Q depends upon R7 which is also controlling
the gain. The dependency of Q and gain could be eliminated by slight
design modification. By adding a 4th amplifier, independent control of
the Q and gain is realized as shown in figure 30. By adding a 4th
amplifier, independent control of the Q and gain is realized as shown
in fig. The State-variable is ideal for high Q circuits [6]. Q's of 500 or
more are easily attainable with a proper filter design [12]. In addition
to high Q, State-variable also gives very high-pass output. Another
feature is that Q and f_m are independent of each other (Tobin, 1998).

Independent the control of Q as well as the gain is thus realized and
also the stability of mid-frequency is also taken care of by State-
variable topology. And unlike the single op-amp architectures, the

open-loop gain (3Q) need only be slightly higher than the filter's output gain (Q), and the low-pass gain is Q, which reduces the requirements on the op-amps GBW (note that for the Sallen-Key, the open-loop gain has to be a minimum of $90Q^2$ which with a Q of 500 would be 22.5MHz) [6].

The fourth amplifier is only required for notch filters. It is also very easy to tune, and it is easy to change the style of low-pass and high-pass, and easy to change the Q of the band-pass and notch.

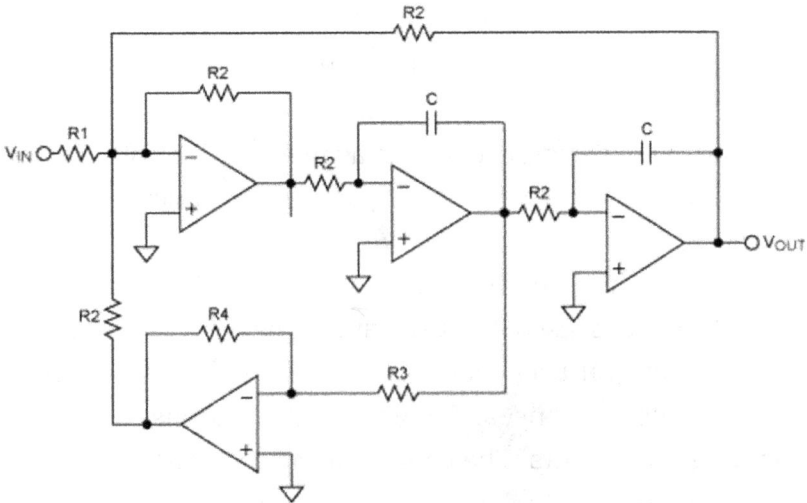

Fig 30: State-variable filter with independent control of Q

Of the topologies discussed thus far, the State-variable is the least sensitive to component variations. It also has another unique attribute: as the frequency f_m changes, the Q and percentage bandwidth remains constant. That is, as we shift f_m in the frequency domain, the Q value remains the same, but the bandwidth of the filter decreases as we increase f_m and increases as we decrease f_m. The percentage bandwidth is defined as [6]:

$$Percentage\ bandwidth = \frac{F_U - F_L}{F_U . F_L} * 100\%$$

Where F_U= upper 3dB bandwidth point

And $\quad F_L$= lower 3dB bandwidth point

And $\quad f_m = F_U F_L$

The major drawback to the State-variable design is the use of 3 or 4 amplifiers which have the negative effect of in power-sensitive applications due to the power supply rejection ratio. However, the effects of adding another dual amplifier much outweigh the cost of special passive components, so it is a viable topology (Maxim, 2002). The design itself is fairly straightforward due to the plethora of filter software and filter design cookbooks available [6]. Another dual op-amp will not cost as much as special passive components will, so it is a viable topology.

Unfortunately, it is not as nice a topology as Akerberg-Mossberg. The same resistor is used for gain and style of filter / Q, limiting control of the filter. It uses more number of amplifiers, but integrate LP, HP and BP filters in one, and any transfer functions can be realized by combining the outputs. The configuration can work very well with a single supply [9].

5.4   Biquad Topology

Lastly, we look at the Biquad filter. It is quite similar to the State-variable Biquad shown above. Biquad is a well-known topology and it is only available in low-pass and band-pass varieties. The low-pass filter is useful whenever simultaneous normal and inverted outputs are needed [7].

It comprises an integrator followed by an inverter and then another integrator, refers to figure 31. Some circuits present it with an integrator and inverter reversed, which makes no difference to response. However, the narrow band-pass output is available at first integrator output. There is just a small change in the circuit organization in terms of individual blocks, but this subtle change provides a circuit that behaves differently than the State-variable filter [6]. In fact, the circuit shown in figure 28 is the actual Biquad circuit derived from the State-variable filter.

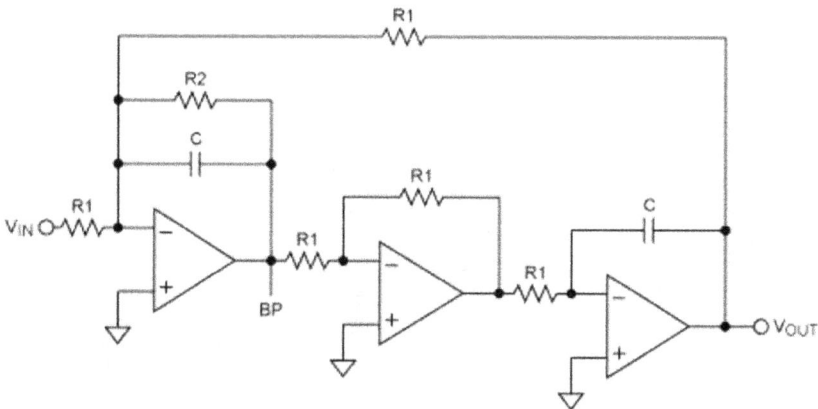

Fig 31: Biquad BPF with Single Supply

The big difference is that for a Biquad, as f_m changes, the bandwidth stays constant, but the Q value changes. Thus, if we change f_m in the frequency domain, as f_m increases, the Q value increases and as f_m decreases, the Q value also decreases. Other than this difference, the Biquad behaves like the State-variable. It allows very high Q values, it can be configured in a 3 or 4 amplifier configuration, and it too is less sensitive to external component variations [6].

It may be observed that it is perfectly all right to change the order of gain amplifier and output integrator. There is another derivative of Biquad, whereby integrator and gain stages are interchanged and gain integrator comes in the middle. This latter configuration is called

Tow Thomos Biquad and textbooks invariably refer the Tow-Thomos (TT) circuit as Biquad. The single supply version of the Biquad is shown in figure 32, which is precisely a TT filter [4]. The performance of both the filter is almost the same. It is all a matter of whether we are summing increased LP signal into an incoming signal at first integrator-cum-summer (as in TT) or we are increasing BP signal available at first integrator-cum-summer to increase its level before extracting LP signal from it and summing it back to the input.

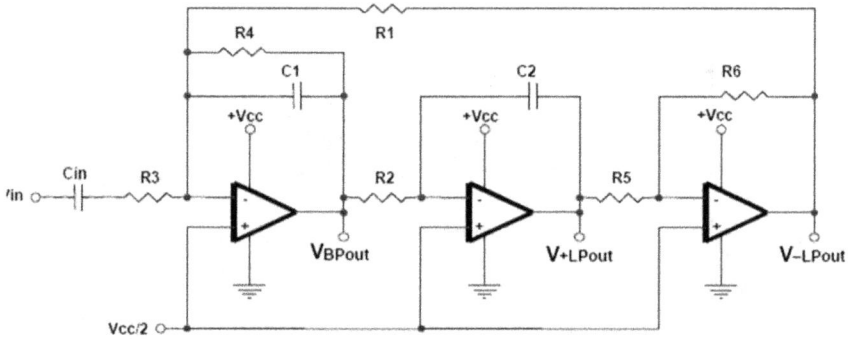

Fig 32: Biquad (TT) BPF with Single supply

However, the single supply TT Biquad circuit is just for illustration. Practically we implemented the actual Biquad circuit only.

For implementation with simpler values of resistors, we choose R1=R2=R5=R and C1=C2=C.

R5 and R6 are not critical; they can be equal resistors. As mentioned, designer can choose any value, as a guideline, R6 is chosen in such a manner that [7]:

$$R6 = 0.707R$$

Keep R3=R4 for unity gain. Otherwise, the absolute gain at mid-frequency will be given by

$$A_m = \frac{R4}{R3}$$

The component selection addressed all the three concerns discussed earlier for successful narrow band filter design.

Independent Mid-Frequency Control

The values of resistor R and capacitor C determine the mid-frequency. Both are related as:

$$f_m = \frac{1}{(2\pi RC)}$$

It is worth noting that R affects f_m, but it does not have much effect on Q. For 577Hz mid –frequency R=274K (nearest E96 resistor to the calculated value of 276K) and C=1nf.

Independent Control of Q:

The Q of the circuit depends upon the value of R4, which need to be high. We chose R4=10M for really good Q value. This simple one resistor Q control makes it such a simple to use topology. The relative ratio of R4 and R determine the Q value of the Q as:

$$Q = \frac{R4}{R}$$

Here it is to be asserted that change in R will not be not very much and once fixed will never be changed. So f_m and Q can be taken independently of each other with R mainly controlling the f_m and R4 controlling the Q value.

Independent Gain Control

As mentioned earlier, the mid-frequency gain is:

$$A_m = \frac{R4}{R3}$$

Generally, the gain is not provided in the Biquad; the gain is controlled mainly by R3. We keep R4=R3= 10M to provide unity gain to the integrator, and R5=R6=10K for unity gain of the gain stage. The actual gain needed is given by an additional gain amplifier, which allows separate gain control independent of Q value.

From the implementation aspect, we found that the 3 op-amp implementation can be done with the help of one quad op-amp. The circuit is quite immune to external component variations. It is worthy to mention that a Biquad has only 2 critical capacitors and 3 resistors inside the Biquad loop. Instead of 2% PPS capacitors NPO 5% capacitors could be used easily.

Frequency Response/Performance

The 3 and 4 amplifier circuits draw more power, and usually require more design time, especially when multiple stages are cascaded to obtain a steeper filter roll-off response. Also, it is costlier as a single amplifier is cheaper than a quad. However, in terms of overall cost of active as well as passive components viz a viz performance, Biquad turns out to be economical.

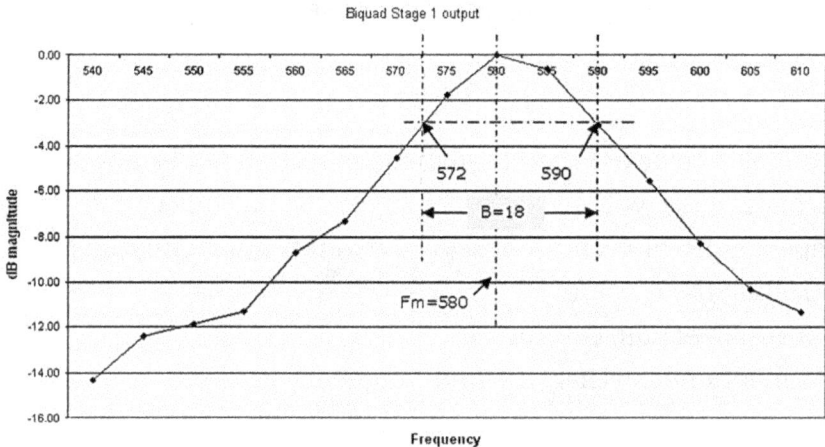

Fig 33: Biquad Stage 1 frequency response

The disadvantage with Biquad is its roll-off may not be steeper enough, but mid-frequency was quite stable at the desired frequency with the desired accuracy. The measured frequency response of one stage Biquad is shown in figure 33. The response gives bandwidth =18 Hz, f_m= 580 (with 274K standard resistor, exact 577Hz can be achieved by using a series 2.2 K resistor), Q= 32. A sharp peak is seen at 577 ± 3 Hz.

Mid-frequency was quite stable at the desired frequency with the desired accuracy. The circuit did not show any oscillations too.

Cascaded Biquad Filter

Now the steeper roll-off could be achieved with the help of cascading two Biquad BPF together; this arrangement gives a very good steep roll-off, much better than other topologies and remarkable stability of mid-frequency, immunity to variation to the external components and relaxed tolerance restrictions on the components. A gain stage may also be inserted in between stage-1 and Stage-2 of the Biquad as shown in figure 34.

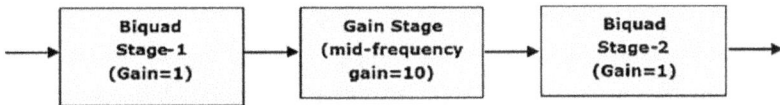

Fig 34: Cascading of Biquad Stages

From a performance point of view, the cascading gives much better results in terms of higher gain (20dB), high Q value (approx. 52), narrower bandwidth (11 Hz) and sharper roll-off (24db /decade) as shown in figure 35. Since the Q curve become sharper due to an extra roll-off and by the unique combination of superimposition of LP and HP signals, the circuit is bound to be stable even at very high Q value.

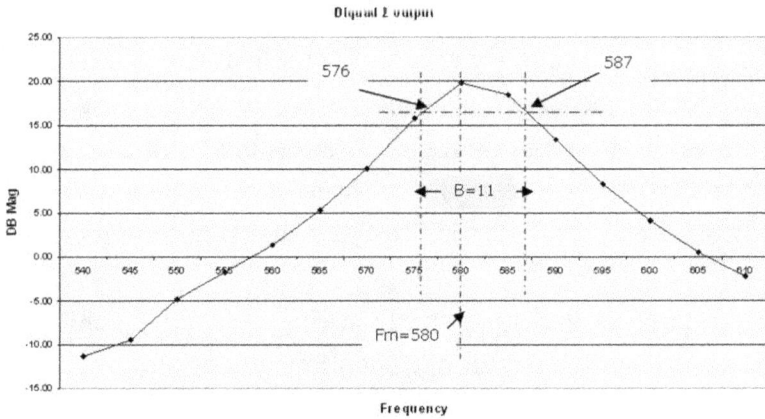

Fig 35: Biquad Stage 2 frequency response

The complete design for a highly stabilized narrow band-pass filter designed as a two-stage Biquad with an independent gain control amplifier is shown in figure 36. The circuit is designed around LMV324 amplifier series with a center frequency of 577Hz and bandwidth of 20Hz.

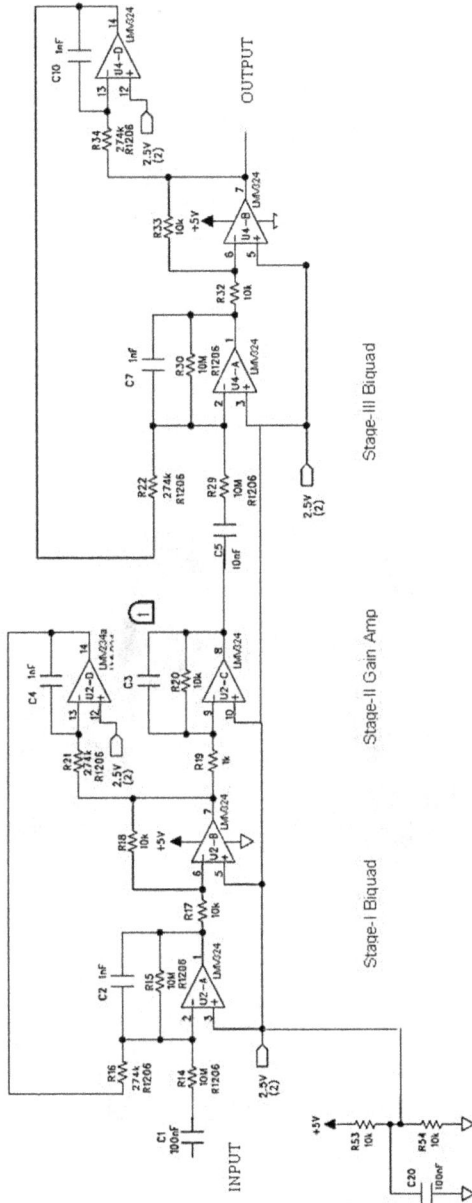

Fig 36: Three stage Biquad filter

The Monte Carlo simulation of the Biquad double-stage circuit containing 7 amplifiers and numerous resistors and capacitors suggests that despite the component tolerance, the bandwidth and the mid-frequency do not get changed much. The maximum variation of mid-frequency is from 571 Hz to 582 Hz, a ± 6 Hz variation with component tolerance, well within the range. The simulation results are shown in figure 37. Further good temperature stability of the mid-frequency is also obtained.

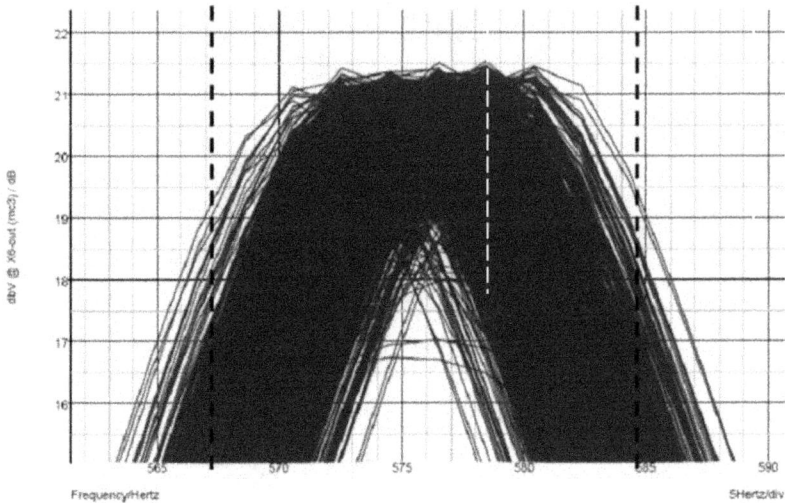

Fig 37: Montecarlo Simulation of Biquad double-stage circuit

In conclusion, the double-stage Biquad filter meets all our set out requirements, provides highly stabilized performance, center frequency at 577HZ with ± 6Hz accuracy, narrow bandwidth of 11Hz, a gain of 20dB, high Q value of 52),) and a sharper roll-off of 24db /decade. Overall independent control on the gain, Q and bandwidth with least temperature or component tolerance variations.

Whereas the other 2-amplifier topologies require some tradeoffs and have inherent dependencies of Q, f_m and gain, the Biquad proved to most efficient in terms of stability of the circuit and immunity to the external components. Biquad multi-stage topology proved very efficient in terms of mid-frequency stability, high Q factor, independent gain and Q values, high roll-off and rejection of power supply harmonics. It is worth noting that multi-stage Biquad filter topology we used has total 7 amplifiers in its circuit and numerous passive components. Despite all this, the Biquad is not that much sensitive to external component variations. This particular fact makes it a very stable topology for the current application. In case of a broad band-pass filter one stage of Biquad is sufficed. However, for narrow band-pass filter two stages may be needed to be cascaded. In total, the circuit performed as per the requirements.

◆ ◆ ◆

CHAPTER 6 – COMPONENT SELECTION CONSIDERATIONS

Theoretically, any values of R and C that satisfy the equations may be used, but practical considerations call for component selection guidelines. Given a specific corner frequency, the values of C and R are inversely proportional—as C is made larger, R becomes smaller and vice versa. Deviations from nominal values of the passive components, of course, influence the frequency response of the filter [8]. These deviations may be caused by component tolerances or because under normal circumstances the ideal values are not available. As a rough estimation, the lower the stage order is, the lower the influence of deviations on the frequency response is. Higher stage orders have a higher quality factor Q, and deviations of R and C impinge on the resulting frequency response roughly proportional to the Q factor. We recommend carrying out Monte Carlo simulation of the desired parameter over all the tolerance ranges of the components involved.

6.1   Selection Considerations for Resistors

A more general rule is that any resistor value in the op-amp RC network should be at least ten times the output resistance of the op-amp and less than one-tenth the input resistance of the op-amp. Secondly, resistors come in various packaging with different precision. It is always recommended to go for 1% resistor tolerances.

These resistors are easily available now. A general overview of types of suitable resistors available for good filter design is given in table 3.

Table 3: Suitability of Various resistors

Type	Temp. Coeff. ppm/deg C	Standard Tolerances %	Comments
Metal film	-25 -->100	1	low cost; most widely used
Cermet film	200	0.5,1	larger, costlier than metal film
Carbon film	-200-->-500	2,5,10,20	low cost; negative temp.coeff.

Metal film resistors are a preference for filter networks. Carbon film resistors may also be used because their negative temperature coefficient can be used to advantage to minimize the passive sensitivity of a circuit (if the capacitors have a positive temperature coefficient). Carbon composition resistors are not suitable for filter networks because of their high-temperature coefficient and high noise level. Carbon composition variable resistors are also not recommended.

Resistor values should stay within the range of 1k ohms to 100k ohms [1]. The lower limit avoids excessive current draw from the op-amp output, which is particularly important for single supply op-amps in power-sensitive applications. Those amplifiers have typical output currents of between 1 mA and 5 mA. At a supply voltage of 5 V, this current translates to a minimum of 1k resistor. The upper limit of 100k ohm is to avoid excessive resistor noise [1]. One could use surface mount components for high heat dissipation and precision characteristics.

To sum up, some points regarding the selection of resistors are [5]:

– Values in the range of a few hundred ohms to a few thousand ohms are best.

– Use metal film with low-temperature coefficients.

– Use 1% tolerance (or better) from E96 series [7].

– Surface mount is preferred.

6.2   Selection Considerations for Capacitors

Capacitors are the real accuracy controlling and variation controlling components. Thus it requires greater attention in their selection, particularly for narrow band-pass filter circuits. These days' capacitors come in various types, the few ones shown in the table 4.

Table 4: Various types of capacitors suitable for filter design

Type	Temp. Coeff. ppm/deg C	Comments
NPO ceramic	0 +/– 30	most popular for active filters
Film: MPC	0 +/– 50	metallized polycarbonate
Film: Polystyrene	–120	larger than MPC; melts at low temp
Mica	–200 ..+ 200	larger, costlier than NPO

NPO (COG) ceramic capacitors are recommended for high-performance filters to minimize the variations of f_m and Q [1]. These capacitors hold their nominal value over a wide temperature and voltage range. The various temperature characteristics of ceramic capacitors are identified by a three-symbol code such as COG, X7R, Z5U, and Y5V. COG-type ceramic capacitors are the most precise. Their nominal values range from 0.5 pF to approximately 47 nF with initial tolerances from ± 0.25 pF for smaller values and up to ±1% for higher values. Their capacitance drift over temperature is typically 30ppm/°C. X7R-type ceramic capacitors range from 100pF to 2.2uF with an initial tolerance of +1% and a capacitance drift over temperature of ±15%. For higher values, tantalum electrolytic capacitors should be used [1].

Other precision capacitors are silver mica, metalized polycarbonate, and for high temperatures, polypropylene or polystyrene. The predictable negative temperature coefficient of polystyrene capacitors can be used to advantage with a metal film or cermet film resistors to minimize passive sensitivity [1].

Standard tolerances all of these capacitors include 1, 2, 5 and 10%. The tolerance of the selected capacitors and resistors depends on the filter sensitivity and the filter performance. Sensitivity is the measure of the vulnerability of a filter's performance to changes in component values. The important filter parameters to consider are the corner frequency, f_m, and Q.

For example, when Q changes by 2% due to a 5% change in the capacitance value, then the sensitivity of Q to capacity changes is expressed as [1]:

$$S\left(\frac{Q}{C}\right) = \frac{2\%}{5\%} = 0.4\frac{\%}{\%}$$

The following sensitivity approximations apply to second-order Sallen-Key and MFB filters:

$$S\left(\frac{Q}{C}\right) \approx S\left(\frac{Q}{R}\right) \approx S\left(\frac{f_c}{C}\right) \approx S\left(\frac{f_c}{R}\right) \approx +0.5\frac{\%}{\%}$$

Although 0.5 %/% is a small difference from the ideal parameter, in the case of higher-order filters, the combination of small Q and f_m differences in each partial filter can significantly modify the overall filter response from its intended characteristic. Figure 38 shows how an intended eighth-order Butterworth low-pass can turn into a low-pass with Tschebyscheff characteristic mainly due to capacitance changes from the partial filters [1]. The difference between ideal and real response peaks with 0.35 dB at approximately 30 kHz, which is equivalent to an enormous 4.1% gain error, can be seen.

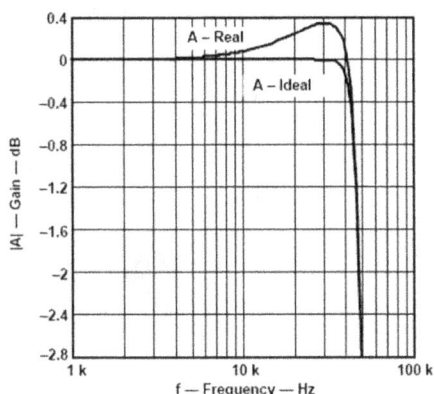

Fig 38: Deviation from ideal response due to change in Capacitor

If this filter is intended for a data acquisition application, it could be used at best in a 4-bit system. In comparison, if the maximum full-scale error of a 12-bit system is given with half least significant bit (LSB), then maximum pass-band deviation would be − 0.001 dB, or 0.012%.

As a general rule, we used 1% tolerance components. 1%, 50V, NPO, SMD, ceramic caps in standard E12 series values are available from various sources. Capacitors with only 5% tolerances should be avoided in critical tuned circuits.

Since capacitor values are not as finely subdivided as resistor values, the capacitor values should be defined before selecting resistors. Capacitor values can range from 1 nF to several uF. The lower limit avoids coming too close to parasitic capacitances [1]. If precision capacitors are not available to provide an accurate filter response, then it is necessary to measure the individual capacitor values and to calculate the resistors accordingly. The capacitor range is chosen depending upon the mid-frequency range. A simple guideline is enumerated in table 5.

Table 5: Mid-frequency vs recommended capacitor values [1]

f_m = Frequency Cut		Capacity in pF	
from	to	from	to
10	100	100000	470000
100	500	22000	100000
500	1000	6800	39000
1000	5000	2700	10000
5000	10000	1000	3300
10000	50000	560	1500
100000	500000	330	1000

Some points regarding capacitor selection include [5]:

- Avoid values less than 100 pF.
- Use NPO if at all possible. X7R is OK in a pinch. Avoid Z5U and other low-quality dielectrics. In critical applications, even higher quality dielectrics like polyester, polycarbonate, Mylar, etc., may be required.
- Use 1% tolerance components. 1%, 50V, NPO, SMD, ceramic caps in standard E12 series values are available from various sources [5, 7].
- Capacitors with only 5% tolerances should be avoided in critical tuned circuits [7].
- Surface mount is preferred.

6.3   Selection Considerations for Operational Amplifiers

The most important op-amp parameter for proper filter functionality is the unity-gain bandwidth. In general, the open-loop gain (AOL) should be 100 times (40 dB above) the peak gain (Q) of a filter section to allow a maximum gain error of 1% [1]. The concept is self-explanatory from figure 39.

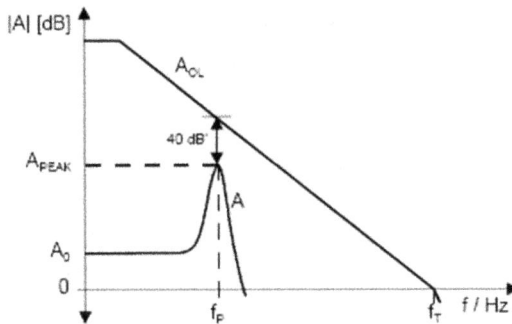

Fig 39: Q value relationship with the amplifier's GBW

The following equations are good rules of thumb to determine the necessary unity-gain bandwidth of an op-amp for an individual filter section [1].

1) First-order filter:

$$f_T = 100. G. f_c$$

2) Second-order filter (Q < 1):

$$f_T = 100. G. f_c. k_i$$

With

$$k_i = \frac{f_{ci}}{f_c}$$

3) Second-order filter (Q > 1):

$$f_T = 100. G. \frac{f_c}{a_i} . \sqrt{\frac{Q_i^2 - 0.5}{Q_i^2 - 0.25}}$$

Besides good DC performance, low noise, and low signal distortion, another important parameter that determines the speed of an op-amp is the slew rate (SR) [1]. For adequate full power response, the slew rate must be greater than:

$$SR = \pi.V_{pp}.f_c$$

A manufacturer's data book will usually include the unity-gain bandwidth in the op-amp's electrical characteristics. Often, a minimum, typical and maximum value will be given, from which the GBW tolerance can be estimated:

$$Tol\ (in\%) = (TYP - MIN)/(TYP)*100$$

The range of op-amp tolerances is wide; from around 15% to about 50%, with the mean approximately 30%. The GBW temperature coefficient must be estimated from a graph of (normalized) unity-gain bandwidth versus free air temperature. This graph is not always provided. Typical coefficient values might be 1000 to 7000 ppm/deg C. Some of the LMV series op-amps from National or Texas Instruments (TI) and TLV series op-amps from TI are good candidates for single supply extremely narrow band-pass filter applications.

To avoid auto-oscillations connect two 100000 pF capacitors between ground and power pins (+V, -V) of op-amp integrated circuit (IC) as shown in figure 40.

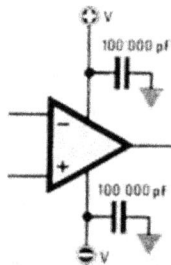

Fig 40: Power supply Decoupling

◆ ◆ ◆

CHAPTER 7 – SOURCES OF NOISE AND REDUCTION TECHNIQUES IN INSTRUMENTATION

An important consideration while designing filters or any other instrumentation system is the amount of noise that may impact the measurement and methods to avoid that noise. This chapter is added to the book to provide general introduction and considerations regarding noise sources which apply to filters discussed in this book.

7.1   Noise Sources

Noise is the biggest environmental factor, which determines if a system will operate reliably in practice. Noise can be random or repetitive, occurring continuously or in an isolated burst. It may affect current or voltage and may occur at any frequency from DC to very high frequencies. A particular source may generate noise over a narrow or wide band of frequencies. Sometimes the effect of noise on the system performance can be very drastic. The overall performance of the circuit is entirely dependent on its noise characteristics. A highly critically designed system may prove to be a failure if it does not conform to the noise characteristics demand by the application. It should be understood that the effect of noise is usually application dependent. Noise fetches the extra amount of attention, especially in the instrumentation and measurements systems. The overall accuracies of such systems is highly affected by the noise and the interference. It is necessary to find out the remedy techniques for noise because it prevails almost in every electronics system.

Once the noise can be observed in a system, it is already a composite signal from many sources. So the first step is to isolate and identify each noise source. With knowledge of mechanisms, which produce noise, design rules can be introduced to ensure that systems are both tolerant to received noise and can meet the statutory restrictions on emitted noise. This paper discusses various practical techniques to remove noise from instrumentation systems.

The strategy for minimizing the effects of noise depends upon the location of its source. The noise problems can generate either in the outer world or it can be caused and communicated within the system, or it can be the local problem with a particular circuit or connection. In the above view noise is classified as below:

a) Internal noise
b) External noise
c) Local noise

7.2   Internal Noise Sources

These noises are generally internal to the system, generated internally into the components of the electronics systems. Following are the major components of the source of such noise.

White Noise

It is uniform noise over entire frequency spectrum and has Gaussian amplitude distribution. It generally evades all the frequencies equally. Two mechanisms are causing white noise.

Shot noise (Shottky Noise)

It is a noise current caused by the fact current flow is not a continuous process but is due to the movements of individual electrons, which are discrete charged particles. The rms value of shot noise over a chosen frequency `f' is given by

$$s = 2qI_bf$$

Where I_b is the bias current flowing and `q' is the charge on the electron.

Thermal Noise (Johnson's noise)

It is due to the random motion of thermally charged particles in any resistive path. The charges will occur randomly at the two ends giving rise to a noise voltage, which increases with temperature. Its rms value over chosen bandwidth is given by

$$V_t = 4\,K\,T\,R\,f$$

Where k is Boltzmann's constant, R is resistance, T absolute temperature. This noise comes in the picture due to resistances used at the source and amplifiers.

Fliker Noise (1/F Noise)

This noise is superimposed on the shot noise in a very low range of frequency. This additional source of noise has been observed both in valves and transistors. This noise increases when the frequency decreases. This is normally called flicker noise or 1/f noise. The flicker noise r.m.s value is given by:

$$V = K \ln (f_h / f_1)$$

Where K= noise content in a decade which has negligible

White noise in 0.1 to 1 Hz F_h is corner frequency which separates the dominant white and pink noise region.

The flicker noise occurs when a localized barrier controls the current in the circuit, whereas shot noise is due to random way electrons surmount the barrier. Flicker noise is due to fluctuations in the effectiveness of the barrier.

Popcorn Noise

These internal noise sources are commonly caused by imperfections in the semiconductor production process or materials. For example, a low-frequency burst bias faults produce a current change in the surface of processed wafers. This is normally called popcorn noise.

Barkhausen Noise

This gets introduced if an instrumentation system employs some magnetic Sensor. This occurs due to the finite size of domains in the ferromagnetic material and the random manner in which directions of orientation of such domains are changed during magnetization

Contact Noise

Contact noise is due to the breakdown of contact between adjacent particles in the path of the current. It manifests itself in the excess noise which occurs in a great many carbon resistors when the current is passed through them. The improved design of the component can greatly reduce such a noise.

Transit Time Noise

If the time taken by an electron to travel from (say) emitter to the collector of a transistor becomes comparable to the period of the signal being amplified, so-called transit time effect takes place and noise input admittance of the transistor increases. This occurs at a frequency in upper VHF range and beyond. Once this high-frequency noise makes its presence felt, it goes on increasing with frequency. at a rate that soon approaches 6dB per octave and this random noise then quickly predominates over the other noise.

Partition Noise

It occurs only in multi-electrode valves which are no more in use. It exists where ultimate destination of any single electron is extremely random because of the availability of two or more electrodes to which an electron can travel.

7.3 External Noise Sources

The external noise is the effects of the external atmosphere or majority effects of other electrical systems. Following are some of the external sources of noise.

Switching Current & Voltages

High current load, which is, switched on/off causes transients in its power supply lines due to their inductance or capacitance. Switching mode power supplied (SMPS), which generates signal noise between 1 KHz and 10 kHz. Switching supplies shows a discrete set of pulses at the fundamental switching frequency or harmonics of it.

Power Lines Interference

Phase control circuits using thyristors and similar devices show a continuous interference spectrum, which is a bit harder to filter. In a high current flowing in a line parallel to the signal line will produce noise it even if it is some distance away. Over only a meter parallel run of lines a meter apart, a mill volt of noise will cause from the current supplying a 3kw load. Also, a fluorescent lighting system produces noise at 100 Hz or 120 Hz depending upon the local main frequency.

Sparking And Radiation

Anything which causes a spark or arching will radiate an electromagnetic wave as noise. Relay and switch arcing and motors with the commentators or slip ring falls into this category. A similar pattern of noise is produced by radio and TV stations which deliberately pollute the ether. Noise in 30 MHz to GHz region gets radiated in the form of an electromagnetic wave. The varying electric and magnetic field produce noise in the other systems. Noise can be radiated at a low frequency of 100 kHz even from the free wires. Long unshielded lines and open conductors.

Environmental And Atmospheric Noise

Spurious radio waves caused by the lightning discharge in thunderstorms and other natural electric disturbances occurring in the atmosphere. It is in the form of impulses and spread over all radio spectrum normally used for the broadcasting. It is less severe in the freq above 30MHz.

Solar noise due to the variability of sun radiations, solar cycle, and sunspots.

Cosmic noise due to thermal or black body noise distributed uniformly over the entire sky by the stars, quasars, and galaxies. Above noise have little effects on instrumentation systems.

Electrostatic Discharge

It is a serious problem in areas with the continental climate or building with an air conditioner and much drier air. A build-up is possible on almost any material. The rate of charging makes the ESD very damaging. The current pulse from an ESD can rise at four amperes per nanosecond.

7.4   Local Noise Sources

This noise is internal to the system but external to some particular set of circuits. It is mainly associated with interconnection of circuits, interfacing of boards, transmission cables, and power supply effects. It should be understood that this type of noise warrants special design attention which shall be discussed later. Following are the sources of local noise.

Coupling Noise

At low frequency, the impedance of concern is either resistive or capacitive, while at higher frequencies the inductance of a short length can become an important factor.

Capacitive coupling

Capacitive coupled interference is associated with the presence of a varying electrostatic field, or difference in potential, between two conductors coupled by some stray capacitance. For example, two pins in a connector can be the `plates of the capacitor, while the mounting insulator is the 'dielectric.'

Magnetic coupling

It is associated with the presence of a varying field in the vicinity of the signal paths. For example, if one of the two pins considered above is conducting a varying current, there exists a varying magnetic field around it. The second pin, if it is the part of an electrically closed circuit, can have a voltage induced by the varying field, resulting in noise.

Current coupling

Current coupled noise occurs where the signal and other currents use a common path, intentionally or otherwise. Any impedance in this path causes interfering currents to develop extraneous signals, which the following device cannot distinguish from the data. For example, if a conductor were used as the 'low' side of the signal path from the transducer to its associated amplifier, and as 'low' side of an AC excitation for another device, the common resistance would develop a component of AC that would appear as a signal.

Crosstalk

Crosstalk is the noise when one signal line picks up the signal from the signal line running parallel to it. Crosstalk is mostly by E-field coupling. Faraday screen (discussed later) is used to reduce it, but the screen efficiency reduces as the frequency of signal increases. Though printed circuit board may exhibit the problem, crosstalk becomes acute when many lines must run in parallel over a significant distance. The main factor affecting the amount of crosstalk are -signal rise time, path length, the geometry of path and termination of the path.

Power Supply Induced Noise

The power supply coupling is caused by the use of the Transformers in the power supply circuits which induces certain inductive coupling in the circuit. Spikes and sags in the currents are some noise evading the system internally.

Interfacing & Cable Noise

This is the noise from interfacing of circuits through cables. The cable or the interconnection wires build up a potential along the length and also get exposed to the external fields. Improper ground loops do create some problems.

7.5   Internal Noise Reduction Techniques

This noises the noise which is the unavoidable noise, and there are some noises which cannot be removed. Since the internal noises are usually uniformly distributed over the entire frequency spectrum, the bandwidth of analog section of a system should be limited to that really needed. Source resistance and the bias currents should be as small as other consideration permits. For VHF range operation RF transistors should be used which are remarkably low noise

7.6   External Noise Reduction Techniques

As discussed above, noise gets radiated from the open & free wires and close running signal wires and the external sources. Capacitive coupling is resulted because of this. This should be minimized by avoiding the free wires and a close run of parallel signal wires. Radiated noise causes some problems related to inductive coupling as explained above.

Faraday Shield

An alternate route may be provided for noise by an earthed shield put between the source and the victim of the noise. It is called a Faraday shield. It restricts E–field propagation. The shield which fully encloses components is called Faraday cage.

Electrostatic Discharge (ESD) Protection

ESD protection can be given by connecting zener diode from line to ground. Connecting two zeners back to back in series gives protection against surges of either sign. Its voltage limiting effect restricts voltage from ESD or electromagnetic EMP to a safe amount by diverting the excess current to ground.

Enclosure Shield

Plastic enclosure shield is best to protect from radiated noise. The enclosure should be continuous & should have a minimum of holes and passages. When a box must have a door, continuity can be retained by using a conductive gasket to give a nearly continuous contact.

Joints & seams should be reduced. According to electrostatic theory conductive shield is the best shield. Now many conductive coatings of an acrylic resin-based paints containing nickel loading with a thickness of 0.025 to 0.05 mm can be done on the plastic enclosures. This conductive layer would give resistivity of fewer than 1.5 ohms per square.

Power lines and high voltage sources should either be kept away from the signal lines or otherwise, both should be properly shielded. The sensitive circuit should not be allowed to work near the fluorescent tube.

The external noise can be reduced mainly by a proper enclosure shielding of the system. Many of the rules of shielding, screening, grounding, and cables as well as the enclosures of power supply are discussed in the next section.

7.7   Local Noise Reduction Techniques

The external noise has been found to become minimum by proper overall shielding, coating and sometimes by proper layout of components. The most severe noise arises due to the local noise which is internal to the system. This topic needs elaborate consideration because it is only noise in the hand of the designer. It is discussed in detail as follows.

Shielding for Capacitive Coupling

The amplifier is particularly susceptible to electrostatic coupling. To reject electrostatic voltages, the input of a charge amplifier must be entirely shielded, with this shield connected to the input common of the amplifier.

Besides shielding a special printed circuit track layout called a Guard is used to surrounds the input pins and is connected to the shield and ground. The track itself forms a shield and prevents any stray currents crossing the surface to the inputs.

In some systems where the interface (connector, for example) in the input cable is being used, input leads must be well shielded in the interface area. It is better to use coaxial cable connectors.

Use of good quality cable is important to stabilize shunting capacitance and minimize leakage. High insulation resistance cable that should enter the receiver chassis using an enclosed connector.

Shielding for Magnetic Coupling

Electromagnetic field sources such as power transformers, solenoids, motors and other electrical devices to be a sufficient distance away that the induced voltage is negligible.

Minimizing signal cable lengths and ensuring wide separations.

Twisting the conductor carrying currents tightly with the return current conductors. This results in the cancellation of outgoing and incoming emf and resultant reduction.

Shielding should be grounded in such a way to reduce the loop area. Grounding the shield at one end only provide electric field shield but no magnetic shield. If the shield is grounded at both the ends, the difference in ground potential at two points may result into a noise current. The rule is to provide a ground at one end at freq less than 1 MHz and ground at both ends above this frequency.

There are certain rules for shielding of cables and the system which are enumerated below for reference:

An electrostatic shield enclosure, to be effective, should be connected to the zero-signal reference potential of any circuitry contained within the shield.

The shield conductor should be connected to zero signal reference potential at the signal earth connection.

When the shield is broken into segments, the shield is required to be tied in tandem as one conductor and then connected to zero signal reference potential at the signal earth point.

The number of separate shields required in a system is equal to the number of independent signals being processed plus one for each power entrance.

Techniques to Reduce Ground Loops

Uniform cable and in particular, employing equal resistors in series with each cable lead as terminations should be used. Otherwise, common-mode signals shall be produced at the output of the input amplifier in case of transducer–amplifier combination

It is essential to provide a low impedance between the cable screen and ground at the transducer terminal or the amplifier terminal. The loop impedance should be made as high as possible

Techniques to Reduce High-frequency Effects

A conductor of a low resistance path at 50Hz can have a significant inductive impedance at 500kHz. In this case, high-frequency source (viz. Computers, transmitters, switching circuits, fluorescent tubes) should be identified, and it should be suppressed at the source itself.

Conducted interference present at high frequencies because of stray capacitance from primary to secondary of the power transformer can be removed by the highly shielded transformer.

One technique of reducing high-frequency ground loop currents consists of increasing the impedance of the cable screen by winding the cable on the ferrite core, thereby creating a coil of greatly increased series inductance.

Well-designed high-frequency amplifier for removal of modulation of the desired signal by unwanted radio frequency noise.

Techniques to Reduce Interconnecting Cable Noise

Triboelectric effect (noise associated with the relative motion, the localized separation between the cable dielectric & outer shield across the dielectric) is removed by the low noise cable having its dielectric coated with a conductive powder to maintain better contact between the outer conductor and the dielectric.

Cable should be securely tied down to avoid motion and bends, and it should be held to a gentle radius to avoid strains.

Techniques to Reduce Power Supply Noise

Use of the main filter which reduces spikes, transients, and RFI by 60 to 65 dB. The simplest filter puts a capacitor in parallel with the supply to act as a low-pass filter.

Adding an inductor (choke) in series with the supply gives a further damping effect by absorbing high-frequency energy in the magnetization core.

Use of torroidal wound inductor with balanced winding connected in both live and neutral lines opposing fluxes in the core. This prevents saturation and gives improved attenuation of the symmetrical noise.

Protection is improved by using a Ferro resonant isolating transformer. This improves sags and surges if supply frequency is constant. The transformer has a tuned resonant secondary winding which produces a current limited, regulated sinusoidal voltage output. It reduces spikes, transients, and RFI and can hold the supply up for about half a cycle in the event of a very short interruption

◆ ◆ ◆

REFERENCES

[1] Thomas Kugelstadt, Active Filter Design techniques: Op-amps for everyone, Texas Instruments www.ti.com, Design Reference SLOD006B, Aug 2002.

[2] Kerry Lacanette, A basic Introduction to filters–Active, Passive and Switched-capacitor, National Semiconductor, Application Note 779, April 1991. http://www.national.com. (Accessed 1 July 2005)

[3] Electronics Circuit Collection – Second order Band-pass Filter Design Topologies, http://www.technick.net/public/code/circuits.php (Accessed 1 July 2005).

[4] Single Supply Analog Expert: On-line Filter design Guide, Texas Instruments, http://www-k.ext.ti.com/SRVS/Data/ti/KnowledgeBases/analog/document/faqs/ssexpert.htm (Accessed 1 July 2005).

[5] James Karki, Analysis of the Sallen-key Architecture, Texas Instruments. Application Report SLOA024A, July 1999. http://www.ti.com (Accessed 1 July 2005).

[6] A Beginners Guide to filter topologies, Maxim, Application Note 1762, Sept 2002.

[7] Bruce Carter, A single Supply op-amp circuit collection, Texas Instruments. Application Report SLOA058, Nov 2000. http://www.ti.com (Accessed 1 July 2005).

[8] Ube Bies, Design and Dimensioning of Active Filters, April 2005. http://www.beis.de/Elektronik/Filter/ActiveLPFilter.html (Accessed 1 July 2005).

[9] Bruce Carter, Filter design in thirty second, Texas Instruments. Application Report SLOA093, Dec 2001. http://www.ti.com (Accessed 1 July 2005).

[10] Active Filter Solutions, Filter Solutions, www.filter-solutions.com (Accessed 1 July 2005).

References

[11] Paul Tobin, Electric Circuit Theory Notes-Chapter 8: State Variable Topology, Dublin Inst of Technology, 1998, http://www.electronics.dit.ie/staff/ptobin/3cover1.pdf (Accessed 1 July 2005).

[12] Maxim Inc, Analog Filter Design Demystified, Maxim Application Note AN 1795. 2002, http://www.maxim-ic.com/appnotes.cfm/an_pk/1795/ (Accessed 1 July 2005)

FURTHER READINGS

1.	Jim Karki, Active Low Pass filter design, Texas Instruments, Application Report SLOA049A, Oct 2000. http://www.ti.com (Accessed 1 July 2005).

2.	Steven Green, Design Notes for 2-pole filter design with differential inputs, Cirrus Logic, Application Note AN48, Mar 2003. http://www.cirrus.com (Accessed 1 July 2005).

3.	Rod Elliott, Multiple Feedback Bandpass filter, Elliott Sound Products, Jul 2000. http://sound.westhost.com (Accessed 1 July 2005).

4.	Analog Filter Design Demystified, Maxim, Application Note AN 1795, April 2002

5.	A Filter Design Primer, Maxim, Application Note 733, Feb, 2001

6.	Alireza Aghashani, State Variable Topology (Second-Order Active Filters Based on the Two-Integrator-Loop Topology), San Jose State University, Filter Design Web Assisted Course EE175. 2000. http://www.engr.sjsu.edu/filter/ (Accessed 1 July 2005).

3.	Green, S. (2003), Design Notes for 2-pole filter design with differential inputs, Cirrus Logic Application Note AN48. http:// www.cirrus.com (Accessed 1 July 2005)

4.	Maxim Inc (2001), A Filter Design Primer, Maxim Application Note 733. http://microblog.routed.net/wp-content/uploads/2006/08/an733.pdf (Accessed 1 July 2005)

FILTER DESIGN SOFTWARE TOOLS

1.	SWIFT ((Switcher with Integrated FET Technology) Designer, Texas Instruments

2.	FilterPro for windows, Texas instruments

3.	MicroCAP 8.0, Spectrum Software

INDEX

THE AUTHOR

Raman K Attri is a corporate business researcher, learning strategist, and management consultant. Masters in electronics engineering, he served as an electronics design scientist at a premier research organization. He has served at technical and product development roles at leading international corporations. As an engineer, he specializes in systems engineering of complex equipment, scientific instrumentation sensors and system design. His international professional career spanned over 25 years across a range of disciplines such as scientific research, systems engineering, management consulting, training operations, and learning design. With his technical and training background, he focuses on the competitive strategies to develop the technical workforce with higher-order troubleshooting and problem-solving skills at a much faster rate. He provides strategic consulting to the organizations by accelerating time-to-proficiency of employees through well-researched models. He holds a doctorate in business from Southern Cross University, Australia.

Speed To Proficiency
RESEARCH

Accelerated Performance for Accelerated Times

Highly-specialized know-how, learning, and resources to solve challenges of 'time' and 'speed' in performance at organizational, professional and personal levels.

Visit us at https://www.speedtoproficiency.com/

S2Pro© Speed To Proficiency Research is a corporate research and consulting forum that provides authentic guidelines to business practitioners to accelerate proficiency of their workforce, teams, and professionals at the 'speed of business'. S2Pro© publishes reports, ebooks, and articles exclusively related to accelerated performance, accelerated proficiency and accelerated expertise in individual and organizational context. Our extensive knowledge base of "how to methods" is derived from experience-based and practice-based observations, analysis/synthesis of existing research, or based on planned/focused research studies through a network of researchers who exclusively focus on 'time' and 'speed' metrics in the business context.

Speed To Proficiency Research: S2Pro©
A research and consulting forum
Singapore 560463

Website: https://www.speedtoproficiency.com
e-mail: rkattri@speedtoproficiency.com
Facebook: https://www.facebook.com/speedtoproficiency/
LinkedIn: https://www.linkedin.com/company/speedtoproficiency/
Twitter: https://www.twitter.com/speed2expertise
Google+: https://plus.google.com/101561704929830160312

www.ingramcontent.com/pod-product-compliance
Lightning Source LLC
Chambersburg PA
CBHW071457210326
41597CB00018B/2584